SPRINGER SERIES IN SURFACE SCIENCES 40

Springer
Berlin
Heidelberg
New York
Barcelona
Hong Kong
London
Milan
Paris
Tokyo

SPRINGER SERIES IN SURFACE SCIENCES

Series Editors: G. Ertl, R. Gomer, H. Lüth and D.L. Mills

This series covers the whole spectrum of surface sciences, including structure and dynamics of clean and adsorbate-covered surfaces, thin films, basic surface effects, analytical methods and also the physics and chemistry of interfaces. Written by leading researchers in the field, the books are intended primarily for researchers in academia and industry and for graduate students.

38 **Progress in Transmission Electron Microscopy 1**
Concepts and Techniques
Editors: X.-F. Zhang, Z. Zhang

39 **Progress in Transmission Electron Microscopy 2**
Applications in Materials Science
Editors: X.-F. Zhang, Z. Zhang

40 **Giant Magneto-Resistance Devices**
By E. Hirota, H. Sakakima, and K. Inomata

41 **The Physics of Ultra-High-Density Magnetic Recording**
Editors: M.L. Plumer, J. van Ek, and D. Weller

Series homepage – http://www.springer.de/phys/books/ssss/

Volumes 1–37 are listed at the end of the book

E. Hirota H. Sakakima K. Inomata

Giant Magneto-Resistance Devices

With 167 Figures

Springer

Professor Dr. Eiichi Hirota
Professor Dr. Hirosi Sakakima
Advanced Technology Reserach Laboratories, Matushita Electric Industrial Co., Ltd.
3-4 Hikaridai, Seika, Souraku, Kyoto, 619-0237, Japan
E-mail: e-hirota@muj.biglobe.ne.jp, sakakima@crl.mei.co.jp

Professor Dr. Koichiro Inomata
Department of Materials Science, Tohoku University
02 Aza-aoba, Aramaki, Aoba-ku, Sendai 980-8579, Japan
E-mail: inomata@material.tohoku.ac.jp

Series Editors:

Professor Dr. Gerhard Ertl
Fritz-Haber-Institute der Max-Planck-Gesellschaft, Faradayweg 4-6,
14195 Berlin, Germany

Professor Robert Gomer, Ph.D.
The James Franck Institute, The University of Chicago, 5640 Ellis Avenue,
Chicago, IL 60637, USA

Professor Dr. Hans Lüth
Institut für Schicht- und Ionentechnik
Forschungszentrum Jülich GmbH,
52425 Jülich, Germany

Professor Douglas L. Mills, Ph.D.
Department of Physics, University of California,
Irvine, CA 92717, USA

Library of Congress Cataloging-in-Publication Data applied for

Die Deutsche Bibliothek – CIP-Einheitsaufnahme

Hirota, Eiichi:
Giant magneto-resistance devices / E. Hirota ; H. Sakakima ; K. Inomata. - Berlin ; Heidelberg ; New York ; Barcelona ; Hong Kong ; London ; Milan ; Paris ; Tokyo : Springer, 2002
(Springer series in surface sciences ; 40)
(Physics and astronomy library)
ISBN 3-540-41819-9

ISSN 0931-5195
ISBN 3-540-41819-9 Springer-Verlag Berlin Heidelberg New York

Springer-Verlag Berlin Heidelberg New York
a member of BertelsmannSpringer Science+Business Media GmbH
http://www.springer.de

Printed in Germany

Typesetting: Layout by author and Marianne Schillinger-Dietrich (Offsetmontagen), Berlin
Cover concept: eStudio Calamar Steinen
Cover production: *design & production* GmbH, Heidelberg

Printed on acid-free paper SPIN: 10779009 57/3020/Rw - 5 4 3 2 1 0

Preface

In the last decade after the discovery of "giant magnetoresistance effects" in metallic multilayers worldwide developments in basic research and in engineering applications have been achieved, and various remarkable results have been obtained in both fields. On the basic research into the GMR effects an excellent review book edited by Hartmann was published in 1999, entitled "Magnetic Multilayers and Giant Magnetoresistance", and it describes the experimental and theoretical aspects of GMR studies and the magneto-optics in metallic multilayers, including applications in electronic data storage fields. This book aims to give an overview on the electronics applications of the GMR in metallic multilayers, especially on the sensors and memories in magnetic data storage, the main concerns of electronic engineers.

If the mean free path of conduction electrons is longer than the period of the multilayer, and for a low applied magnetic field magnetizations in the adjacent magnetic layers are in antiparallel, the GMR effect in magnetic metallic multilayers occurs due to the resistance change in the electric current flowing in the layer plain, with a change in the magnetization from antiparallel to parallel alignment by an applied magnetic field (CIP mode), i.e., in the saturation field magnetizations are in parallel and the resistance is lowered. The saturation field H_s is given by the RKKY-like magnetic coupling between adjacent magnetic layers through a nonmagnetic conductive layer between the layers. In early GMR multilayers H_s was larger than 0.1 T and they could not be used in electronic devices. Electronic applications need multilayer systems with low H_s, such as the [NiFe/Cu/Co/Cu] multilayer with a H_s of 1 mT. These low-H_s systems and their sensor applications are described. After proposal of "spin valve" or "pseudo-spin-valve" techniques for novel devices, which are essentially trilayer devices composed of a nonmagnetic layer sandwiched between two magnetic layers, the development research shifts to these spin-valve-type devices which are now used in the market in read heads in hard disk drives (HDDs). This is one of the main subjects of this book.

When an electric current flows perpendicular to the layer plain (CPP mode), the GMR effect becomes larger than that in the CIP mode. But devices in this mode show impedances too low for use. A practical GMR device in CPP mode is constructed by using a ferromagnetic tunnel junction (MTJ), i.e., the nonmagnetic conductive layer in the spin valve is replaced by an insulator film. This so called TMR device has high potential for future electronics applications. A probable application is magnetic random access memory (MRAM) and this is another main subject in this book. The forecast for other possible spin electronics devices is also given in this book.

The purpose of this book is to provide a text on GMR devices for electronic-material scientists and/or device engineers, and for graduate students in these fields, and give an overview of applications and elemental physics models related to GMR and TMR devices. The authors' backgrounds are in solid-state physics where we accumulated knowledge and experience through research and development into electronic materials and devices, especially from magnetic materials and related devices.

This book started with a recommendation from Dr. S. Yosimura and he introduced us to his friend Dr. Claus E. Ascheron of Springer-Verlag. We would like to express our thanks for their continuous concern and support.

Osaka, Kyoto and Sendai, Japan
May 2001

E. Hirota
H. Sakakima
K. Inomata

Contents

1 Introduction

The discovery of the giant magnetoresistance (GMR) effect in metal multilayers make a strong impact on scientific researchs and also the applied sciences, especially on the development of computer memory technologies. The former is to research the origin of this extraordinary large MR effect in artificial materials, such as, metal multilayers. The latter is the possibility of achieving high-density memory in hard disk drives (HDD) of computer peripherals. At present the introduction of GMR read heads in the HDD opens the door to the "Gb/in^2 memory era" and allows aerial recording densities (ARD) of up to 100 Gb/in^2.

In 1971 Hunt [1] proposed using a magnetoresistance transducer, i.e., MR read heads and sensors. After his proposal various studies on MR transducers were carried out aiming to develop practical devices. Early devices were composed with anisotropic ferromagnetic thin films, such as $Fe_{20}Ni_{80}$ alloy film, which have a magnetoresistance (MR) ratio of about 2 % and which were later called anisotropic magnetoresistance (AMR) films as distinct from GMR films. Several AMR devices were put on the market as magnetometric positioning sensors, rotary encoders, and read heads of digital audio tape recorders etc.

By the beginning of the 1990 inductive thin film heads and thin film MR read heads had been put on the HDD market, successively, and attained an ARD of 1 Gb/in^2 (see Fig. 1.5).

Under these circumstances the GMR effect, which shows a MR ratio larger than 10 %, is very attractive as an application of read heads to HDD. Intensive studies of this application have been done in USA, Europe and Japan. Now, GMR heads of 50 Gb/in^2 have been announced and in the near future are sure to be put on the market.

Successful results in GMR devices act as an incentive to study MR effects in ferromagnetic tunnel junctions (TMR), which has a MR ratio larger than 20 %. The TMR head is expected to be the next generation of read heads of HDD with an ARD of over 100 Gb/in^2. The most promising TMR device is the magnetic random access memory (MRAM) and many studies are now progressing with the aim of developing random access memories with high-frequency access, are nonvolatile, have low power consumption and are radiation resistant.

GMR and TMR devices have a basic structure of two magnetic layers separated by a nonmagnetic spacer layer. These effects are the change in electric resistance of this basic structure, when the magnetizations of both magnetic layers change from a parallel configuration to an antiparallel configuration and vice versa, followed with an applied magnetic field. Thus, the key technologies are to fabricate an antiparallel configuration, which is easily converted to a parallel configuration by a small applied field. There are two different methods: one is to use the spin-pinning method and the other is to use magnetic layers with different coercive forces.

The spin-pinning method is to pin the magnetization of one layer by adding an antiferromagnetic layer on it, while the magnetization of another layer rotates freely with applied field. This type of device is called a "spin valve". If two magnetic layers have different coercive forces, a parallel configuration and an antiparallel configuration is created at relatively small field. This type of device is called a "pseudo-spin-valve". These two types of spin valves are used in various devices for their simple structures and low working fields.

Various magnetic memories in the solid state have been investigated over the last 50 years in the research for nonvolatile memories in electronics. Ferromagnetic thin film memory and magnetic bubble memory, etc., have been studied widely but have not been successfully put on the market.

In the middle of the 1980s AMR films were applied to MRAM [2] but AMR-memory cells show low output caused by the low MR ratio of AMR films, and for this reason AMR cells failed to make practical MRAM. In 1995 applications of the GMR effect to MRAM were announced by using a spin valve or a pseudo-spin-valve [3] and[4]. Recent advances in research on the GMR and TMR effect place one's hope in the development of a practical MRAM and extensive studies have been carried out throughout the world.

This chapter is an introduction to the GMR and TMR devices treated in this book. Brief explanations of magnetic transducers, i.e., MR sensors and MR read heads, are given in Sect.1.1 and of the MRAM in Sect.1.2.

Chapter 2 gives the theory of the GMR and TMR effect for a basic understanding of these devices. First is the phenomenological theory and also the micromagnetic theory of spin pinning. Second is the theory of electric conduction in metal thin films on the basis of the Boltzmann transport equation and then the GMR effect is discussed by considering spin-dependent conductivity in the spin-valve structure. Third is the theory of electron tunneling through a potential barrier and the tunnel current is expressed as functions of barrier height and width for various potentials. The TMR effect in a ferromagnetic tunnel junction with a rectangular potential is discussed on the basis of spin-dependent tunneling in the spin-valve structure.

Chapter 3 treats the GMR effect in metal multilayers and its application for sensors of high sensitivity. The GMR effect occurs when the antiparallel initial configuration of adjacent ferromagnetic layers due to the RKKY-like interlayer coupling changes to the parallel configuration with increasing applied field. Therefore, a larger field is necessary than the equivalent field of the interlayer coupling, i.e., the saturation field H_s. A practical GMR device needs a low working field, therefore main concerns are how to realize devices with a low working field. Detailed discussions are given on various structures of the multilayers composed of soft magnetic materials for achieving low H_s.

Chapter 4 discusses the GMR effect in spin valves, which give basically highly sensitive devices working at a low field. A standard spin-valve device consists basically of three layers: a free magnetization layer, a nonmagnetic spacer layer and pinned magnetization layer. Spin pinning is done by adding an antiferromagnetic layer on it or by using a synthetic antiferromagnetic (SAF) layer. The SAF layer is composed of a pinned magnetic layer, a conductive nonmagnetic spacer layer and another magnetic layer, in which the thickness of the spacer layer is adjusted so as to couple both magnetic layers antiferromagnetically, because of the magnetic coupling between both magnetic layers is the RKKY-type coupling via the spacer layer.

A pseudo-spin-valve is a useful device. These spin-valve devices have been used for make devices having the high MR ratio and low working field. In particular, detailed explanations are given of the enhancement of the MR ratio by the addition of a highly conductive back layer and/or of a specular reflective capping layer. The HDD read head and the nondestructive MRAM are explained as applications of GMR devices of spin-valve type.

Chapter 5 and Chap. 6 deal with TMR, i.e., spin-dependent tunneling in a ferromagnetic tunnel junction (MTJ) and its application to MRAM. Starting from basic features of the TMR and microfabrication technology of MTJ cell, the architectures and characteristics of MRAM are discussed, together with the design and performance of MTJ cells. The present and future states of the art are also discussed.

Chapter 7 gives the prospects for TMR devices, i.e., bipolar spin transistor, spin FET, spin-valve transistor, spin-dependent single-electron tunneling devices, devices using the Coulomb blockade effect and the spin accumulation effect, etc.

1.1 Magnetic Transducers (Sensor and Read Head)

Figure 1.1 shows the basic concept of digital magnetic recording of a HDD, in which a magnetic disk moves with relative velocity $\boldsymbol{v}$ to a magnetic head. The head consists of a magnetic core with a narrow gap and n turns of a winding coil. Digital signals due to the write current J in the coil are recorded as residual magnetization patterns on the recording tracks of the disk. The unit of written signals is a bit, of length l_b (bit length) and width T_w (track width). The space between adjacent tracks is T_s and the distance between adjacent tracks T_p (= $T_w + T_s$) is called the track pitch. The head also reads written signals as the induced voltage V in the coil and $V = nfm_rT_wl_bt$, where m_r is residual magnetization of the written signals and $f = v/l_b$; t is the thickness of the magnetic medium on the disk. The magnetic disk is composes of a solid substrate of light metal, such as Al, and a thin recording medium of hard magnetic materials deposited on the substrate.

The performance of the HDD is expressed by the recording densities. There are three kinds of expressions, i.e., the linear recording density $1/l_b$ in bits/in (BPI), the track density $1/T_p$ in tracks/in (TPI) and the aerial recording density $1/l_bT_p$ in bits/in^2 (ARD). For achieving a high recording density it is desirable to have as short a value of l_b as possible. The value of the l_b is determined by leakage flux of the write head and is determined by the gap length g, the write current J and the head/medium spacing h. Signals of shorter l_b than g can be recorded adjusting J or h , but signals of shorter l_b than g cannot be read by the inductive type head. This is a problem for read heads for high-density recording but at present its solution is to use a MR type read head.

High-density recording is also attained by increasing the relative speed $\boldsymbol{v}$ but high speed causes mechanical abrasion of the head and medium. Also, for high-density recording the recording medium must have a high coercive force H_c to keep the recorded signals stably and a high magnetization $\boldsymbol{m}$ to obtain the necessary output to the read head. But these are beyond the scope of this book.

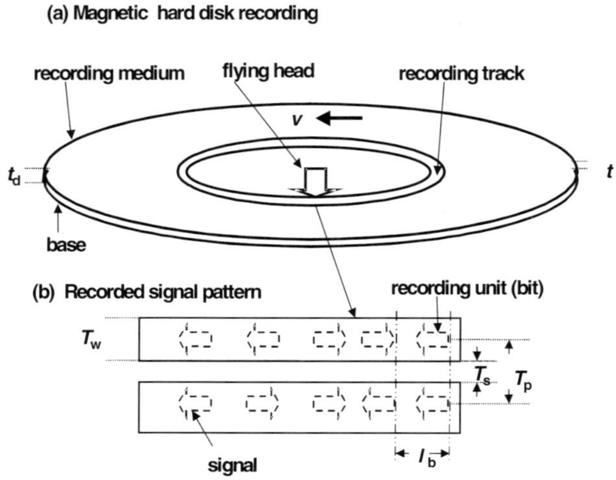

Fig. 1.1. Magnetic disk recording: (**a**) recording method, (**b**) recorded pattern on magnetic disk. l_b = bit length, T_w = track width, T_s = track space, T_p = track pitch ($T_w + T_s$), t = thickness of magnetic medium, t_d = thickness of disk

1.1.1 MR Sensor

A sensor, such as a Hall effect device, was used in early magnetometric system, which measure the linear or rotational movement by sensing a magnetic scale attached to a moving body, but later the need for high resolution required two terminal devices, such as the magnetic read heads used in magnetic recording.

The MR sensor consists of a thin AMR film of rectangular shape and two contacts attached at both longitudinal sides of the film for flowing the sensing current $\boldsymbol{J}$. The magnetic field $\boldsymbol{H}$ of the magnetic scale is perpendicular to $\boldsymbol{J}$ and tilts the magnetization $\boldsymbol{M}$ away from $\boldsymbol{J}$ by an angle $\phi(H)$- ϕ_0. The magnetoresistance (AMR) is

$$R(H) = R_\perp + \Delta R \cos^2[\phi (H)-\phi_0]. \tag{1.1}$$

Here $\Delta R = R_{||} - R_\perp$ with longitudinal resistance $R_{||}$ and perpendicular resistances $R_\perp$, respectively, and the ratio $\chi(= \Delta R/R_{||})$ is called the MR ratio. The usual AMR film consists of a ferromagnetic alloy and its MR ratio χ is about 2~4%. The output voltage of the MR sensor is $V_{AMR} = JR(H)$ and is independent of the relative speed v. The resolution of the sensor is essentially determined by the thickness w of the sensor film. These characterisitcs of the MR sensor are desirable for high-resolution sensors.

To obtain a linear response on $\boldsymbol{H}$ the initial ϕ_0 ($\boldsymbol{H} = 0$) is set up as $\pi/4$. For this purpose a perpendicular bias $\boldsymbol{H}_{b\perp}$ is applied to the AMR film by using a suitable method, for example, by the soft magnetic adjacent layer (SMAL) method as shown in Fig. 1.2. In the reading process the AMR sensor shows large Barkhausen noise caused by magnetic wall movements in the AMR film. To suppress this noise, a longitudinal bias $\boldsymbol{H}_{b||}$ is also applied by piling antiferromagnetic film on the AMR film or by attaching magnets on both sides of the film. Various structures for biasing methods were invented during 1970–1980. Figure 1.2 shows an example of an AMR sensor, in which $\boldsymbol{H}_{b\perp}$ is applied by the SMAL method and $\boldsymbol{H}_{b||}$ is applied by using magnets. In the SMAL method the SMAL is magnetized to $\boldsymbol{M}_{SMAL}$ by the sense current $\boldsymbol{J}$ and $\boldsymbol{M}_{SMAL}$ produces $\boldsymbol{H}_{b\perp}$ in the AMR film.

At present AMR sensors are widely used in various applications, such as encoders for precise rotation control of high-speed motor and precise positioning sensors, etc. Recently, the high-sensitivity GMR sensor, which consists of metal multilayers, was put on the market and details are explained in Chapter 3.

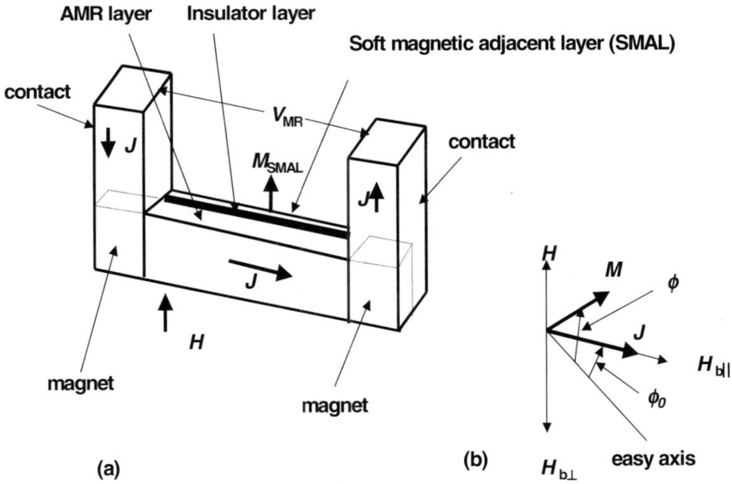

Fig. 1.2. (**a**) Magnetoresistance read head (AMR) with longitudinal bias $\boldsymbol{H}_{b||}$ by magnets and with perpendicular bias $\boldsymbol{H}_{b\perp}$ by soft magnetic adjacent layer (SMAL). (**b**)Relations among the sense current $\boldsymbol{J}$, magnetization $\boldsymbol{M}$ and signal field $\boldsymbol{H}$ in the AMR layer

1.1.2 HDD Head

The read head of the HDD is the most important application of MR devices. The basic structure of the HDD is shown in Fig. 1.1, a hard disk (magnetic recording medium) rotates with high speed $\boldsymbol{v}$ around its center and the head writes and reads digital signals "0" or "1" in circular tracks on the disk. The HDD head is called a flying head because in working the head floats over the fast rotating magnetic disk and keeps aerodynamically a definite space (flying height h) between the head and the disk.

Figure 1.3 shows the working flying head assembly. The write and read (W/R) head is formed on the front side of a slider, which receives a lifting force from rapid air flow and an opposite force from the flexible suspension, and in equilibrium between two forces flies over the disk stably to keep a constant flying height h for W/R head. This slider is called an air bearing slider (ABS). In the advanced HDD higher-density recording needs a smaller h of 100–10 nm and thus one of the key problems for HDD is the design and manufacture the ABS. In Fig. 1.3(a) an example of the ABS surface is shown; this is called a pico-slider and corresponds to an h of about 100 nm.

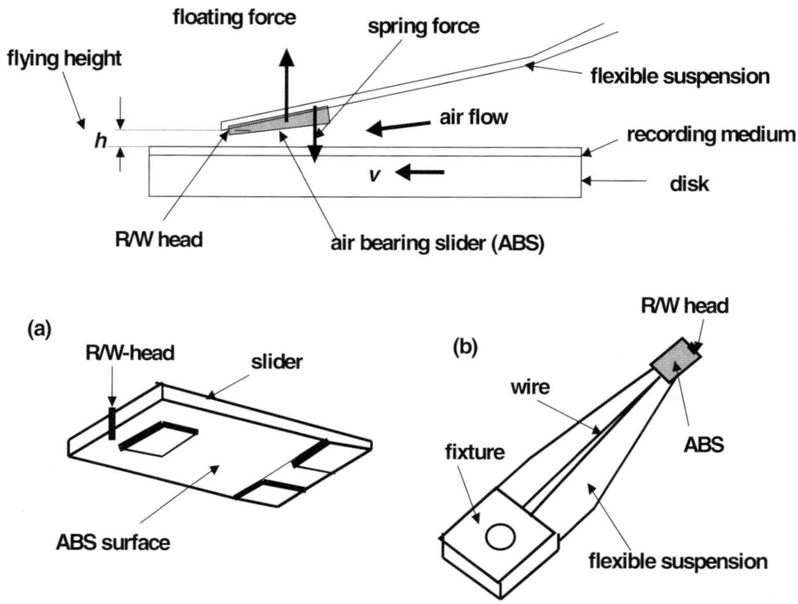

Fig. 1.3. Flying head assembly: (**a**) air bearing slider (ABS), (**b**) flexible suspension

Usually, a flying head works in so-called contact start and stop mode, i.e., the head initially rests on the disk surface and in working takes off to a definite attitude h. After working it again lands on the disk surface. The protection of heads and recorded signals on the disk from mechanical damage in working is a most important problem and therefore numerous studies have been carried out for making extremely flat both the ABS and the recording medium surface and also for finding suitable surface lubricants. Figure 1.3(b) shows the suspension schematically. Recent HDDs need very precise head positioning on the disk and therefore there have been various studies on the suspension composed of a piezoelectric ceramics as an actuator for fine positioning and a preamplifier loaded on it for the signal of the read head.

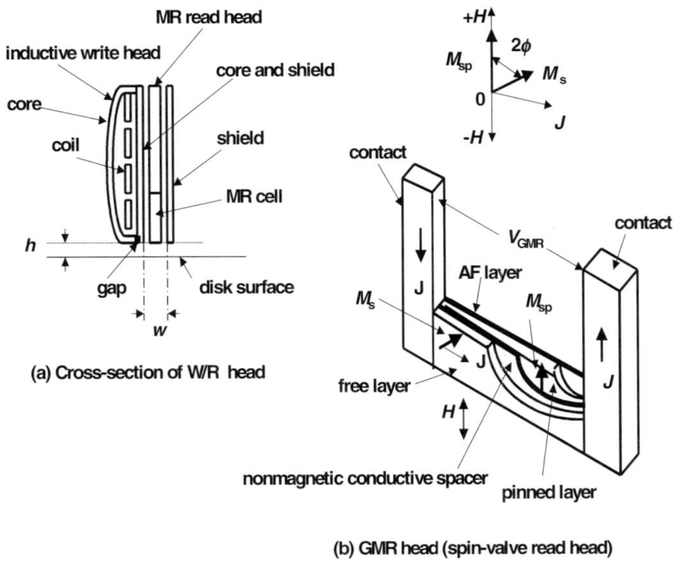

Fig. 1.4. (**a**) HDD write/read combination head; flying height h, effective head gap w; (**b**) GMR read head in spin-valve type: signal field $\boldsymbol{H}$, sensing current $\boldsymbol{J}$

At present, the write head is a thin film inductive head and the read head is a thin film MR head. Both heads are combined into a W/R head and formed on the top of the slider. A cross-sectional view of the combined W/R head is shown in Fig. 1.4(a). Two magnetic shield layers are at both sides of MR read head and their separation is w, which determines the resolution of MR head. In Fig. 1.4(b) GMR head (spin-valve type) is shown schematically, in which two ferromagnetic layers are separated by a conductive nonmagnetic layer and the magnetization $\boldsymbol{M}_{sp}$ of one ferromagnetic layer is pinned by the adjacent antiferromagnetic (AF) layer but the magnetization $\boldsymbol{M}_s$ of another layer rotates freely by the magnetic field $\boldsymbol{H}$ from recorded signals on the disk.

If we denote the angle between $\boldsymbol{M}_s$ and $\boldsymbol{M}_{sp}$ by $2\phi(\mathbf{H})$ and the sense current by $\boldsymbol{J}$, the voltage V_{GMR} between both contacts is given by

$$V_{GMR} = J\,[R(\phi{=}0) + \Delta R \sin^2\phi] \tag{1.2}$$

here $\Delta R{=}R(\phi{=}\pi/2){-}R(\phi{=}0)$ and the ratio $\chi[=\Delta R/R(\phi{=}0)]$ shows the magnetoresistance ratio of GMR head.

The evolution of the HDD expressed by their recording density (ARD) is shown in Fig. 1.5 in which the solid line indicates the ARD of the market HDD and the dotted line indicates laboratory data for the ARD. After the introduction of AMR heads in 1991 the annual evolution rate of the ARD has been higher than it was before. A still higher rate will be achieved by using the GMR head as shown in this figure. The GMR (spin-valve) head achieves an ARD of 10–50 Gb/in^2. With these advances in the HDD the changing trends in the related parameters l_b, T_s and h are shown as functions of the ARD in Fig. 1.6. A high-performance HDD of 10–100 Gb/in^2 requires these parameters to be 100–1 nm. Table 1.1 shows typical characteristics of these high-performance HDD with GMR heads. Recent advances in HDD are reported by Coughlin [5].

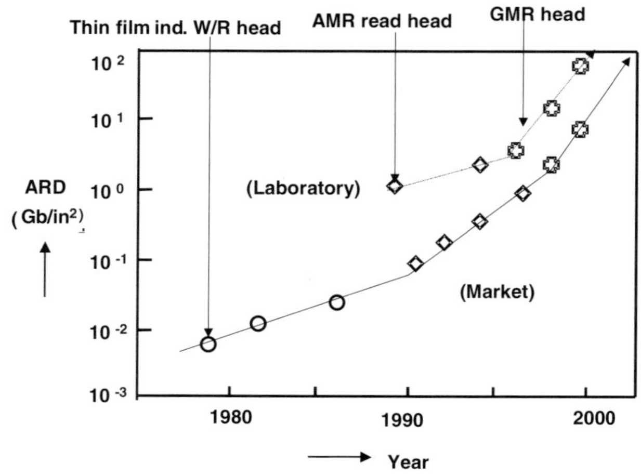

Fig. 1.5. ARD evolution in HDD

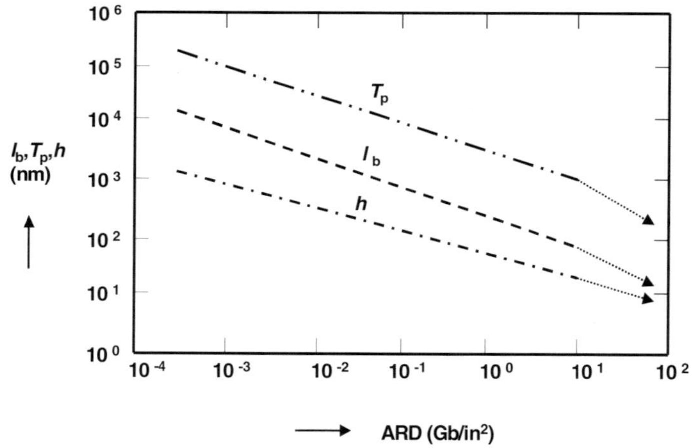

Fig. 1.6. Changing trends in the parameters l_b, T_p, and h as functions of the ARD; bit length l_b, track pitch T_p, flying height h

In the future TMR heads will be used in HDDs of 100 Gb/in^2. To fabricate such high-performance HDDs another key technology is a high-performance disk having an H_c of several thousand kOe together with high residual magnetization, and also having an extremely smooth surface of roughness $t_{p\text{-}p}$(<10 nm). Also, there is the problem so-called of the paramagnetic limitation for thermal stability of recorded signals.

Table 1.1. Characteristics of high-performance HDDs with GMR read head

HDD	ARD (Gb/in^2)	20.3	35.3	56.1
	Linear density (kBPI)	492	522	678
	Track density (kTPI)	41.4	67.7	82.7
	Flying height (nm)	27	23	21
Head	Out put (mV$_{p\text{-}p}$)	1.0	0.4	–
	Write T_w (μm)	0.47	0.37	0.28
	Read T_w (μm)	0.42	0.29	0.21
Maker		IBM	IBM	Fujitu

1.2 Magnetic Random Access Memory (MRAM)

MRAM is constructed with a matrix of bit lines (x-axis) and word lines (y-axis) and with magnetic memory cells formed at each cross point of the matrix wirings. The basic structures of the GMR and the TMR memory cells are shown in Fig. 1.7, in which

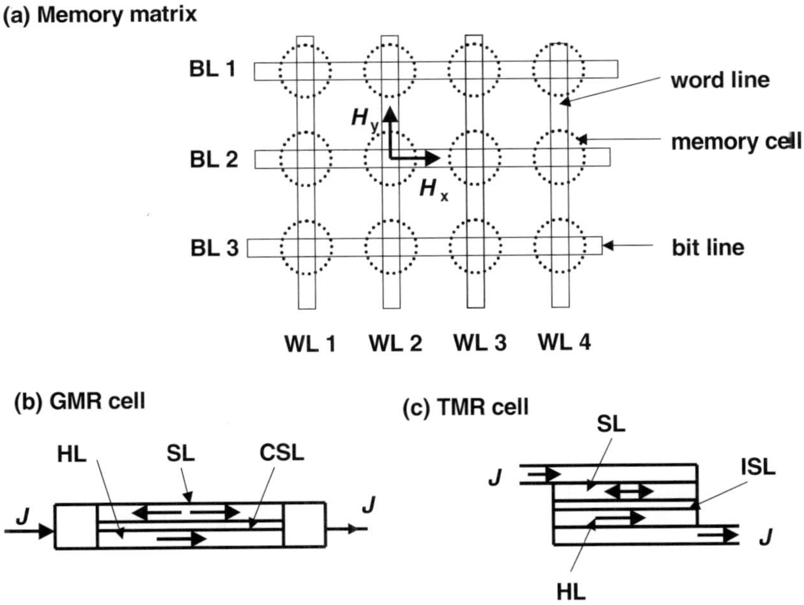

Fig. 1.7. MRAM using GMR or TMR memory cell with a pseud-spin-valve. HL: hard magnetic layer, SL: soft magnetic layer, CSL: conductive space layer, ISL: insulating space layer, J: sensing current

pseudo-spin-valve type cells are used, and a hard magnetic layer (HL) and a soft magnetic layer (SL) compose the sensing cells of the recorded signals. The HL works as a memory cell, which has uniaxial magnetic anisotropy and its easy axis is parallel to the x-axis. Digital signals "0" and "1" are recorded as the residual magnetization $+M_r$ or $-M_r$ of the HL by the signal magnetic field $\boldsymbol{H}$ (H_x, H_y) caused by the currents in the word and bit lines. If we assume a rotational magnetization process in the HL, the magnetization $\boldsymbol{M}$ is switched from $+M_r$ to $-M_r$, or vice versa, when the magnetic fields H_x and H_y satisfy the following condition:

$$H_x^{2/3} + H_y^{2/3} \geq H_k^{2/3}, \tag{1.3}$$

where $H_k=2K/M_s$ with uniaxial anisotropy constant K. Writing or rewriting of memory signals is done by adjusting the current so as to satisfy this condition (the current coincidence method). The various constructions and sensing methods are proposed and detail discussion is given in Chapter 5.

The MRAM and competitive memories are listed in Table 1.2. The ferroelectric random access memory (FERAM) is real nonvolatile memory, which is fabricated by so-called silicon technology, but this memory is difficult to fabricate using ferroelectric materials such as $BaTiO_3$ in semiconductor processes. MRAM and FERAM are now under development and expected performances of these memories are listed in the table.

Table 1.2. Competitive nonvolatile RAM

Memory	MRAM	FERAM	FLASH
Access time (ns)	300(GMR)	~200(GMR)	50
	<60(TMR)	~40(TMR)	
Write time (ns)	<10	$\sim 10^2$	$\sim 10^4$
Repetition cycle	$>10^{15}$	10^{12}	10^5
Capacity (bit/Tip)	>1 Gb	<10 Mb	>1 Gb
Power (mW)	<10	>10	~100

References

[1] R Hunt, *IEEE Trans. Mag.* **MAG-7**, 150 (1971)

[2] A M Pohm, J M Dauguton, C S Comstock, H Y Yoo, and J Hur, *IEEE Trans. Mag.*, **23**, 2575 (1987)

[3] Y Irie, H Sakakima, M Satomi, and Y Kawawake, *J. Appl. Phys.*, **34**. L414 (1995)

[4] D D Tan, P K Wang, V S Sperious, S. Le, and K K Kung, *IEEE Trans. Mag.*, **31**, 3206 (1995)

[5] T M Coughlin, *J. Mag. Soc. Japan*, **25**, 111 (2001)

2 Physics of GMR and TMR Devices

GMR and TMR devices have a basic common structure, namely, two ferromagnetic metal films separated magnetically by a nonmagnetic film. The difference between the structures of these devices is in the nonmagnetic spacer film which consists of a metal film (GMR) or an insulator film (TMR).

The electric resistance R of these structures depends on the relative angle 2ϕ between the magnetizations of both ferromagnetic films. If an applied magnetic field $\boldsymbol{H}$ causes a changes in 2ϕ, the resistance R changes as functions of $\boldsymbol{H}$ and thus shows the magnetoresistance (MR) effect. We indicate by $R_{\uparrow\uparrow}$ and $R_{\uparrow\downarrow}$ the resistances R for two magnetizations in parallel (2ϕ=0) and in antiparallel (2ϕ=π), respectively. Generally, $R_{\uparrow\uparrow} < R_{\uparrow\downarrow}$ and the magnetoresistance ratio χ is given by

$$\chi = (R_{\uparrow\downarrow} - R_{\uparrow\uparrow})/R_{\uparrow\uparrow} . \tag{2.1}$$

The magnetic response of these structures to $\boldsymbol{H}$ is expressed by a hysteresis loop. At sufficiently large $\boldsymbol{H}$ two magnetizations are in parallel and at small negative $\boldsymbol{H}$ two magnetizations become antiparallel, if the structures satisfy the following conditions:

(a) the two ferromagnetic films have different coercive forces (pseudo-spin-valve case);
(b) the magnetization of one ferromagnetic layer is pinned by exchange coupling with antiferromagnetic overlayer and the magnetization of the other layer changes freely with $\boldsymbol{H}$ (spin-valve case);
(c) the pinned layer in the spin-valve is the synthetic antiferromagnetic layer (SAL), which consists of two ferromagnetic films separated by a conductive nonmagnetic film and the magnetizations of both ferromagnetic films couple antiferromagnetically by RKKY-like interlayer exchange coupling via the nonmagnetic conductive spacer layer and the outer magnetic layer pinned by the antiferromagnetic overlayer.

In cases (a) and (b) it is desirable that two ferromagnetic metal layers have a rectangular hysteresis loop with a small coercive force H_c. If $\boldsymbol{H}$ causes a magnetization reversal from antiparallel to parallel and vice versa, a large MR effect occurs in these basic structures (GMR or TMR).

In this chapter the physics of GMR and TMR devices is introduced for the fundamental understanding of these devices. Three topics are treated: first is spin pinning by interface exchange coupling in spin-valve devices (Sect. 2.1); second is the theory of the GMR effect on the basis of spin-dependent conduction in the basic structure (Sect. 2.2); and third is the theory of the TMR effect on the basis of spin-dependent electron tunneling in the magnetic tunnel junction (Sect. 2.3). The interlayer exchange coupling in the SAL structure has the same origin as the interlayer coupling in the metal multilayer and is discussed in Chapter 3.

2.1 Spin Pinning

A typical GMR device is a spin valve, which has excellent performance with high MR ratio with a small working field ***H***. The structural features of a spin valve are shown in Fig.1.4(b). A spin valve is composed of four metal layers: a free magnetic layer (FL); a nonmagnetic conductive spacer layer (SL); a pinned magnetic layer (PL); and an antiferromagnetic layer (AL). Field cooling treatment pins the magnetization of the PL along the field cooling direction by the AL overlayer, while that of the FL rotates freely, following the applied magnetic field ***H***. The spin pinning of the PL is due to unidirectional anisotropy induced by domains in the AL and by interface exchange coupling between the PL and the AL.

The phenomenological theory of this interface coupled system is given in Sect. 2.1.1 in accordance with Fulcomer *et al.* [1]. The micromagnetic theory of spin pinning is given by Koon [2], who examined the spin arrangement on the PL and the AL near their interface by micromagnetic similation methods, and his theory is surveyed in Sect. 2.1.2. These discussions give a clear picture of spin pinning. In this book the atomic magnetic moment is called spin, for short.

2.1.1 Phenomenological Theory of Spin Pinning

In early work on magnetic small particles it was found that a surface oxidized Co particle shifts its hysteresis loop along the direction of the field-cooled treatment. This behavior is called exchange anisotropy, and is caused by the exchange coupling between Co particle and the overlayer of antiferromagnetic CoO.

Spin pinning of the present device has the same origin and we discuss the magnetic behavior of exchange coupled ferro/antiferromagnetic layers. The magnetic hysteresis loop of the PL shifts along the field-cooling direction by $-H_D$, which is a measure of spin pinning.

This shift field H_D depends on the temperature and frequency of the hysteresis measurement. With increasing temperature H_D decreases to zero near the blocking temperature T_B, which is defined below and is lower than the Neel temperature T_N of the AL.

We assume the AL is an assembly of N identical antiferromagnetic particles, which are magnetically independent of each other and has the shape of a thin platelet. As shown in Fig. 2.1 the easy axis (i.e., the field-cooled direction) of the PL is parallel to the x-axis and the magnetization $\boldsymbol{M}$ of the PL tilts towards the x-axis by ϕ. An antiferromagnetic particle has uniaxial anisotropy K with its easy axis tilting towrds the x-axis by φ and has an interface magnetic moment $\boldsymbol{m}$ which tilts towrds the x-axis by θ. The interface exchange interaction between $\boldsymbol{m}$ and $\boldsymbol{M}$ is $J(\boldsymbol{M}\cdot\boldsymbol{m})ac$, here a is the contact area of the particle, ac is the effective area for the exchange coupling. The thickness of particle is d.

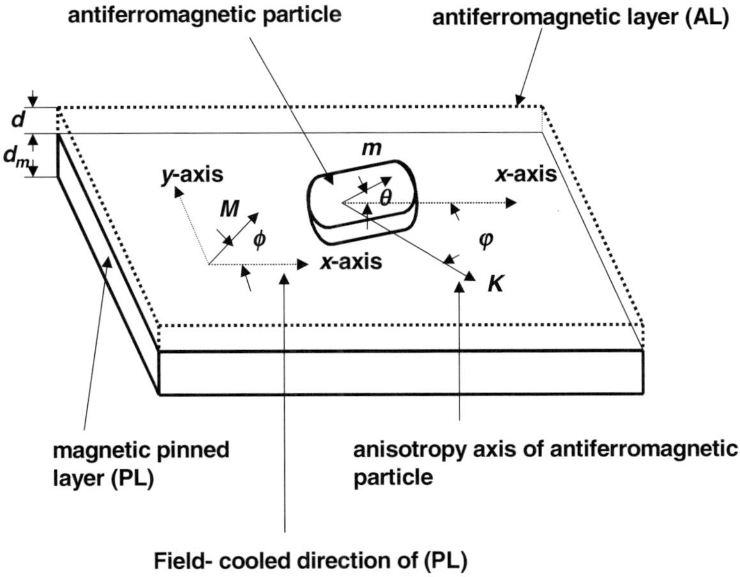

Fig. 2.1. Model of spin pinning by an antiferromagnetic layer composed of the same particles

The magnetic energy of the particle is

$$E_{AF} = Kad \sin^2(\theta\text{-}\varphi)\text{-}J(\boldsymbol{M}\cdot\boldsymbol{m})ac \cos(\theta\text{-}\phi). \tag{2.2}$$

When the field $\boldsymbol{H}$ is applied along the x-axis and changes from $+H_s$ to $-H_s$ (H_s is the saturation field), $\boldsymbol{M}$ inverts to $-\boldsymbol{M}$, namely, ϕ changes from 0 to π. Accordingly the particle energy increases to

$$E_{AF} = Kad \sin^2(\theta\text{-}\varphi) + J(\boldsymbol{M}\cdot\boldsymbol{m})ac \cos \theta. \tag{2.3}$$

To relax this energy increment the particle would like to turn its moment from $\boldsymbol{m}$ to $-\boldsymbol{m}$. At a temperature T and changing frequency f of $\boldsymbol{H}$ we consider this relaxation process assuming φ=0, i.e., the easy axes of the PL and of the AL coincide. This assumption doesn't essentially affect the description of the relaxation process.

We express the energies by their equivalent temperatures, i.e., $e=E_{AF}/\mathrm{k}$, $e_a= Kad/\mathrm{k}$, $e_x=JMmac/\mathrm{k}$ with Boltzmann constant k, (2.3) becomes

$$e = e_a\sin^2 \theta + e_x\cos \theta. \tag{2.4}$$

The relaxation process is described by the change in θ from 0 to π, and midway in this process there is an energy barrier, of height and position determined by the condition $\partial e/\partial\theta$ =0. The barrier position is $\theta_b=\cos^{-1}h$ with $h=e_x/2e_a$ and the height is $e_{max}=e_a(1+h^2)$ as shown in Fig. 2.2. In the case of $h\geq1$(the strong interface coupling case), θ_b=0 and there is no barrier in the relaxation process. $\boldsymbol{m}$ follows the change of $\boldsymbol{M}$. But in the case h<1 there is a barrier and the changing process of $\boldsymbol{m}$ becomes a relaxation type.

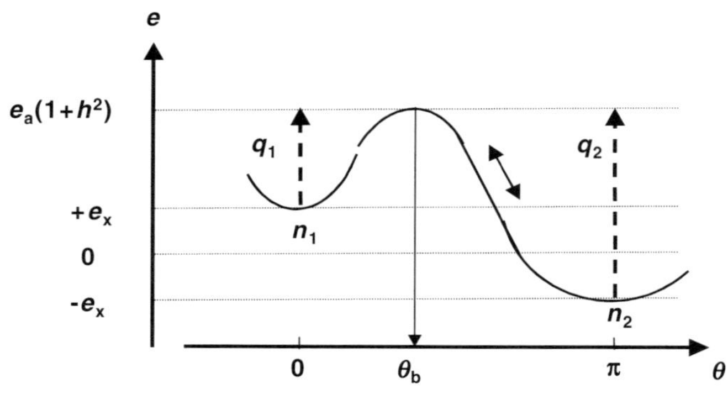

$q_1 = e_a(1+h^2) - e_x$, $q_2 = e_a(1+h^2) + e_x$

$\theta_b = \cos^{-1} h$

Fig. 2.2. Schematic presentation of relaxation process

At a temperature T the changing rate of N_1 (the number of particles at θ=0) is given by

$$dN_1/dt = \nu_0[N_2\exp(-q_2/T) - N_1\exp(-q_1/T)], \tag{2.5}$$

here $\nu_0=\gamma H_a$ (γ:gyro-magnetic ratio, $H_a=2K/m$) and is 2.8×10^9 H_a (Hz/kG), $q_1=e_a(1+h^2)-e_x$, $q_2=e_a(1+h^2)+e_x$,and $N_2=N-N_1$ (the number of particles at $\theta=\pi$).

Putting $n_1=N_1/N$, (2.5) becomes

$$(1/\nu_0)(dn_1/dt) = \exp(-q_2/T) - n_1[\exp(-q_2/T)+\exp(-q_1/T)]. \tag{2.6}$$

Using the relaxation time τ, the solution is given by

$$\{n_1(t)-n_{1\infty}\}/(n_{10}-n_{1\infty}) = \exp(-t/\tau), \tag{2.7}$$

where $n_1(0)=n_{10}$, $n_1(\infty)=n_{1\infty}=1/[1+\exp(-2e_x/T)]$ and the relaxation frequency is $\nu=\nu_0[\exp(-q_1/T)+\exp(-q_2/T)]$.

The average value of $\boldsymbol{m}$ is

$$m(t) = n_1(t)-n_2(t) = 2n_1(t) - 1, \tag{2.8}$$

where we assume the magnitude of $\boldsymbol{m}$ is 1. In thermal equilibrium $m_{1\infty}$ is

$$m_{1\infty} = 2n_{1\infty} - 1 = \tanh(e_x/T), \tag{2.9}$$

which is determined by e_x and T , while the relaxation frequency ν_0 is a function of e_x, e_a and T as follows

$$\nu = 2\nu_0\cosh(e_x/T)\exp[-e_a(1+h^2)/T]. \tag{2.10}$$

In Fig. 2.3 $m_{1\infty}$ versus T is plotted as a function of e_x. $m_{1\infty}$ is nearly 1 at low temperature, and becomes 0.76 at $T=e_x$ and at sufficiently high temperature it tends to zero. At high temperature the relaxation frequency ν approaches ν_0, i.e. m_1 immediately becomes $m_{1\infty}$, while at low temperature the relaxation frequency decreases rapidly to zero and m_1 stays at m_{10}. The temperature at which ν is equal to the changing rate f of $\boldsymbol{H}$ is called the blocking temperature T_B. In hysteresis loop measurements at a lower temperature than T_B, $\boldsymbol{m}$ can't follow $\boldsymbol{H}$, i.e., $\boldsymbol{M}$ is pinned by $\boldsymbol{m}$ at a lower temperature than T_B.

In Fig. 2.4 contour lines of the blocking temperature T_B are shown as functions of e_a and e_x. These lines are calculated by (2.10) with assuming f = 2.8 Hz. The left domain of the 0_+ K line is essentially no-blocking ($2e_a<e_x$), i.e., $\boldsymbol{m}$ follows $\boldsymbol{M}$ and the pinning effect disappears. If the measuring temperature T is 50 K, in the domain between 0 K and 50 K lines, $\boldsymbol{m}$ behaves superparamagnetically and H_D depends on T and f, while in the right domain of the 50 K line the relaxation frequency ν becomes smaller than 2.8 Hz, and $\boldsymbol{M}$ is pinned by $\boldsymbol{m}$ and the hysteresis loop shifts by H_D (spin pinning).

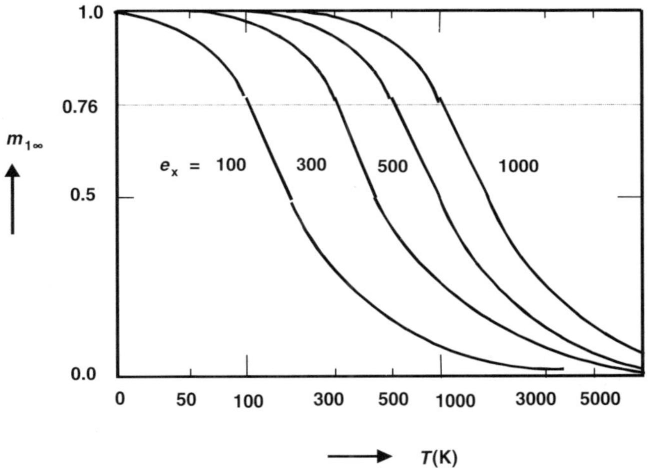

Fig. 2.3. Temperature dependence of equilibrium magnetization $m_{1\infty}$

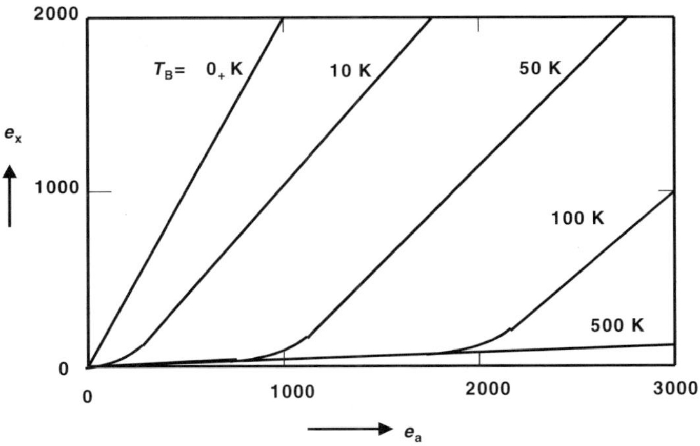

Fig. 2.4. Blocking temperature as functions of exchange coupling e_x and anisotropy energy e_a

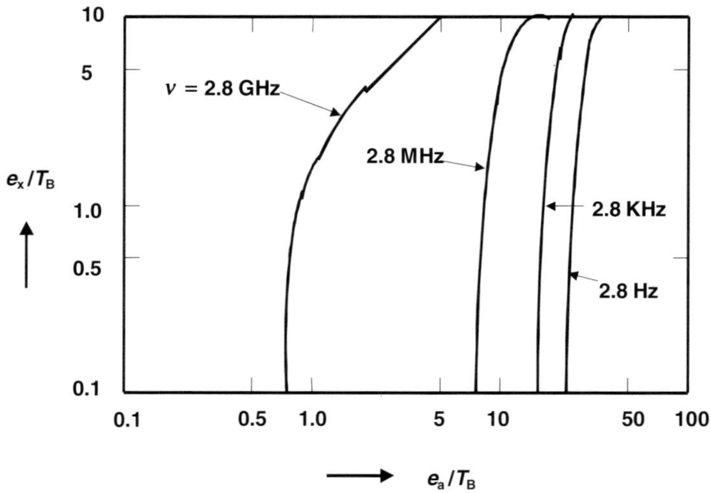

Fig. 2.5. Relaxation frequency as functions of e_g/T_B and e_x/T_B

In Fig. 2.5 another presentation of (2.10) is shown, i.e., the relaxation frequency ν is plotted as functions of e_a/T_B and e_x/T_B. If the measuring frequency f is 2.8 MHz, the left domain of the 2.8 MHz line shows no spin pinning due to more rapid relaxation than 2.8 MHz, but the right domain shows spin pinning. ν depends strongly on e_a/T_B but weakly on e_x/T_B.

Generally, antiferromagnetic particles differ from each other and therefore particles are numbered by j to distinguish them from other particles. In an applied field $\boldsymbol{H}$ the total magnetic energy E_T is expressed as

$$E_T = (MV)[(K_u/M)\sin^2\phi - H_x\cos\phi - H_y\sin\phi] + \sum_{j=1}^{N} (mv_j)[(K/m)\sin^2\theta_j - JM(c_j/d_j)\cos(\theta_j-\phi)]. \quad (2.11)$$

The first term is the energy of the PL with its volume V and uniaxial anisotropy constant K_u, and H_x and H_y are components of $\boldsymbol{H}$. The second term is the energy of the AL, which consists of N particles distinguished by j. This equation can be written as follows:

$$e_T = E_T/MV = (H_k/2)\sin^2\phi - H_x\cos\phi - H_y\sin\phi + \sum_{j=1}^{N} [(H_a^j/2)\sin^2\theta_j - H_e^j\cos(\theta_j-\phi)], \quad (2.12)$$

where $H_k=2K_u/M$, $H_a^j=(2K/m)(mv_j/MV)=2e_{aj}/MV$, and $H_e^j=Ja_jc_j/V=e_j/V$.

In a rotational magnetization process the hysteresis loop is determined by the torque balance of $\partial e_T/\partial\phi=-L_\phi=0$. If we apply $\boldsymbol{H}$ along the x-axis ($H_y=0$), i.e., along the easy axis of the PL, this condition becomes

$$\partial e_T/\partial\phi = (H_k/2)\sin 2\phi + \sin\phi(H_x + \sum_{j=1}^{N} H_e^j\cos\theta_j) - \cos\phi \sum_{j=1}^{N} H_e^j\sin\theta_j = 0. \quad (2.13)$$

From this equation the hysteresis loop is deduced by determing $\cos\varphi(=M_x)$ as functions of H_x, and also the coercive force H_C is determined by the condition $\partial^2 e_T/\partial\phi^2 = 0$. The calculated result is $H_C=-(H_k\cos^3\phi+\sum_{j=1\cdot N} H_{ex}^{\,j}\cos\theta_j)$ with ϕ=0 or π, and the H_C is written as

$$H_{C\pm} = -(\pm H_k)-H_D, \tag{2.14}$$

with $\cos\theta_j=m_j(t)$ and

$$H_D=\sum_{j=1}^{N} H_e^j m_j(t), \tag{2.15}$$

This shows that the hysteresis loop shifts by $-H_D$ (the shift field). In the case of J>0 (ferromagnetic interface coupling), H_D is positive and the hysteresis loop shifts in the negative direction, and in the case of J<0 the loop shifts in the positive direction.

For further calculation of $H_{C\pm}$ and H_D the distribution function $G(e_a,e_x)$ is necessary. $G(e_a,e_x)de_a de_x$ is the fraction of particles, i.e., e_a between e_a and e_a+de_a and e_x between e_x and e_x+de_x, and is normalized as $\int_0^\infty de_a\int_0^\infty de_x\, G(e_a,e_x)= 1$.

Using this function, the shift field is given by

$$H_D= (N/MV)\int_0^\infty de_a\int_0^\infty de_x\, m(t)e_x G(e_a,e_x). \tag{2.16}$$

The maximum shift field H_{Dmax} occurs in perfect pinning, i.e., m(t)= 1, and is given by

$$H_{Dmax}= (N/MV)\int_0^\infty de_a\int_0^\infty de_x\, e_x G(e_a,e_x). \tag{2.17}$$

The energies e_a and e_x are functions of a, c and d, Fulcomer *et al.* [1] calculated $H_{C\pm}$ and H_{Dmax} with starting from specific distribution functions of these parameters and determining $G(e_a,e_x)$ as the joint distribution of these specific distributions. They assumed that there is a maximum, e.g., a_{max} and the distribution function $f(a)=1/a_{max}$ for $a<a_{max}$ and $f(a)=0$ for $a>a_{max}$, therefore $e_{a\,max}= Ka_{max}d_{max}/k$ and $e_{x\,max}= Ja_{max}c_{max}/k$. Using this distribution the maximum shift field becomes

$$H_{Dmax}= \eta Jc_{max}/2M_s d_m, \tag{2.18}$$

where $\eta(=A_a/A)$ is the ratio of the total contact area A_a of particles in the AL to the interface area A of the PL, which are given by $A_a=N\int_0^\infty af(a)da$ and $A=V/d_n$. H_{Dmax} is inversely proportional to the thickness d_m of the PL.

2.1.2 Micromagnetic Theory of Spin Pinning

In the preceding Sect. 2.1.1 the spin arrangement on the AL was assumed to be ferromagnetic at its interface with the PL (in this section this interface is written as the A/F interface and the ferromagnetic pinned layer PL is denoted by FL). But the spin arrangement on the interface layer isn't only ferromagnetic but also antiferromagnetic. Koon [2] studied, in a general way based on the micromagnetic treatment, the spin-pinning effect for any spin arrangements of antiferromagnetic (fully compensated spin arrangement), ferrimagnetic (partially compensated case) or ferromagnetic (uncompensated case) materials. In uncompensated or partially compensated cases spin-pinning effects are clear and essentially the same as in

Sect. 2.1.1 but in the fully compensated case (**m**=0 case) the spin-pinning effect is unclear but can be discussed by following Koon's procedure.

A typical ferromagnetic metal is Fe which has a bct crystal structure and therefore this section treats the spin-pinning effect in a bct crystal, i.e., both the ferromagnetic and antiferromagnetic layers have the bct crystal structure, in which the A/F interface is the (1 1 0) or (1 0 0) crystal plane. If exchange coupling between nearest neighbor atoms in the AL is antiferromagnetic, the spin arrangement on the (1 1 0) plane of the AL is fully compensated and is discussed in Sect. 2.1.2.1 and the spin arrangement on the (1 0 0) plane is ferromagnetic (uncompensated), which is described in Sect. 2.1.2.2.

2.1.2.1 Fully Compensated Case: (1 1 0)-Interface Plane

The unit cell of a bct-crystal lattice is shown in Fig. 2.6(a), in which the two magnetic sublattices are indicated by i(=1,2). The spins on each sublattice are denoted as $\boldsymbol{S}_i$ and have eight nearest neighbor spins which belong the another sublattices. The spacing distance of (1 1 0) plane is $a/\sqrt{2}$ and the density of spins is $\sqrt{2}/ac$, here a and c are lattice parameters of the bct lattice. The exchange energy is $E_{ex}= -2J(\boldsymbol{S}_1 \cdot \boldsymbol{S}_2)$ and $\boldsymbol{S}_1$, $\boldsymbol{S}_2$ lie in (1 1 0) planes and the uniaxial anisotropy energy of the AL is $E_a=-K(S_{iz}/S_i)^2$ with K>0. The easy axis is [0 0 1] direction and is indicated as the z-axis in Fig. 2.6(a). The layer structure near the A/F interface is shown schematically in Fig. 2.6(b). The structure is composed of L of the AL and N of the FL. Layers are numbered from the outside layer of the AL by j (=1,2,...L, L+1,L+2,...L+N) and so the A/F interface layers are the L and L+1 layers, respectively.

Total magnetic energy per unit area of the layer is given by

$$\begin{aligned}
E = & -2J_{AA}\{\boldsymbol{S}_1^{\,1}\cdot(2\boldsymbol{S}_2^{\,1}+\boldsymbol{S}_2^{\,2})+\boldsymbol{S}_2^{\,1}\cdot\boldsymbol{S}_1^{\,2}\}-K\{(S_{1z}^{\,1})^2+(S_{2z}^{\,1})^2\} \\
& -\sum_{j=2}^{L-1} [2J_{AA}\{\boldsymbol{S}_1^{\,j}\cdot(2\boldsymbol{S}_2^{\,j}+\boldsymbol{S}_2^{\,j+1}+\boldsymbol{S}_2^{\,j-1})+\boldsymbol{S}_2^{\,j}\cdot(\boldsymbol{S}_1^{\,j+1}+\boldsymbol{S}_1^{\,j-1})\}+K\{(S_{1z}^{\,j})^2+(S_{2z}^{\,j})^2\}] \\
& -2J_{AA}\{\boldsymbol{S}_1^{\,L}\cdot(2\boldsymbol{S}_2^{\,L}+\boldsymbol{S}_2^{\,L-1})+\boldsymbol{S}_2^{\,L}\cdot\boldsymbol{S}_1^{\,L-1}\} \\
& -2J_{AF}(\boldsymbol{S}_1^{\,L}\cdot\boldsymbol{S}_2^{\,L+1}+\boldsymbol{S}_2^{\,L}\cdot\boldsymbol{S}_1^{\,L+1})-K\{(S_{1z}^{\,L})^2+(S_{2z}^{\,L})^2\} \\
& -2J_{FF}\{\boldsymbol{S}_1^{\,L+1}\cdot(2\boldsymbol{S}_2^{\,L+1}+\boldsymbol{S}_2^{\,L+2})+\boldsymbol{S}_2^{\,L+1}\cdot\boldsymbol{S}_1^{\,L+2}\} \\
& -\sum_{j=L+2}^{N-1} [2J_{FF}\{\boldsymbol{S}_1^{\,j}\cdot(2\boldsymbol{S}_2^{\,j}+\boldsymbol{S}_2^{\,j+1}+\boldsymbol{S}_2^{\,j-1})+\boldsymbol{S}_2^{\,j}\cdot(\boldsymbol{S}_1^{\,j+1}+\boldsymbol{S}_1^{\,j-1})\} \\
& -2J_{FF}\{\boldsymbol{S}_1^{\,N}\cdot(2\boldsymbol{S}_2^{\,N}+\boldsymbol{S}_2^{\,N-1})+\boldsymbol{S}_2^{\,N}\cdot\boldsymbol{S}_1^{\,N-1}\} -\sum_{j=1}^{L+N} \{\boldsymbol{H}\cdot(\boldsymbol{S}_1^{\,j}+\boldsymbol{S}_2^{\,j})\},
\end{aligned} \tag{2.19}$$

where $(2/ac)J=J_{FF}$, J_{AF} and J_{AA} for the FL, the A/F interface and the AL, respectively, and the last term is the Zeeman energy of the applied field $\boldsymbol{H}$.

Using effective magnetic fields $\boldsymbol{H}_{\mathrm{eff}\,1,2}^{\,j}$ this energy can be written as follows:

$$E = -\sum_{j=1}^{L+N} (\boldsymbol{H}_{\mathrm{eff}\,1}^{\,j}\cdot\boldsymbol{S}_1^{\,j}+\boldsymbol{H}_{\mathrm{eff}\,2}^{\,j}\cdot\boldsymbol{S}_2^{\,j}), \tag{2.20}$$

where $\boldsymbol{H}_{\mathrm{eff}\,1,2}^{\,j}=-\partial E/\partial \boldsymbol{S}_{1,2}^{\,j}$. The effective fields are listed in Table 2.1.

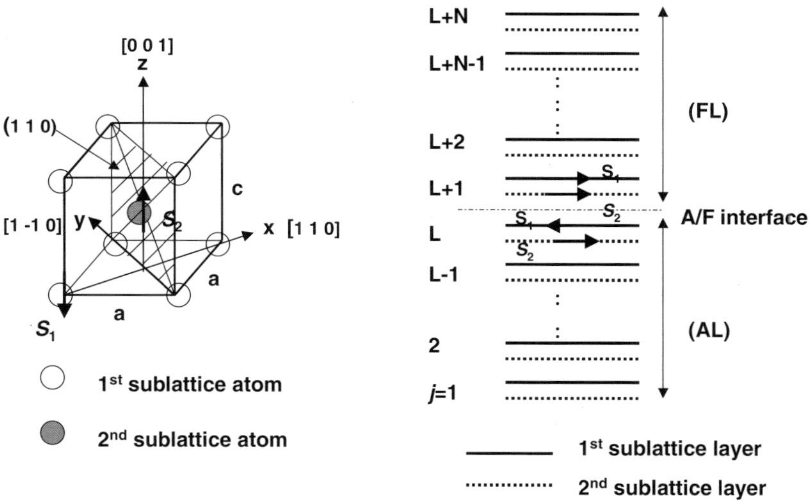

(a) Body-centered tetragonal lattice **(b) (1 1 0) layer structure near A/F interface**

Fig. 2.6. (**a**) Unit cell of bct crystal with two sublattice, and (**b**) schematic presentation of spin arrangement on (1 1 0) layers near the A/F interface

Table 2.1. Effective magnetic fields for $\boldsymbol{S}_{1,2}^{\ j}$ on the (1 1 0) plane for $J_{FF}(>0)$, $J_{AA}(<0)$

layer	$\boldsymbol{S}_{1,2}^{\ j}$	$\boldsymbol{H}_{\text{eff }1,2}^{\ j}$
(A)	$\boldsymbol{S}_{1,2}^{\ 1}$	$2J_{AA}(2\boldsymbol{S}_{2,1}^{\ 1}+\boldsymbol{S}_{2,1}^{\ 2})+2K\boldsymbol{S}_{1,2z}^{\ 1}+\boldsymbol{H}$
(A)	$\boldsymbol{S}_{1,2}^{\ j=2\to L-1}$	$2J_{AA}(2\boldsymbol{S}_{2,1}^{\ j}+\boldsymbol{S}_{2,1}^{\ j+1}+\boldsymbol{S}_{2,1}^{\ j-1})+2K\boldsymbol{S}_{1,2z}^{\ j}+\boldsymbol{H}$
(A)	$\boldsymbol{S}_{1,2}^{\ L}$	$2J_{AA}(2\boldsymbol{S}_{2,1}^{\ L}+\boldsymbol{S}_{2,1}^{\ L-1})+2J_{AF}\boldsymbol{S}_{2,1}^{\ L+1}+2K\boldsymbol{S}_{1,2z}^{\ L}-\boldsymbol{H}$
(F)	$\boldsymbol{S}_{1,2}^{\ L+1}$	$2J_{FF}(2\boldsymbol{S}_{2,1}^{\ L+1}+\boldsymbol{S}_{2,1}^{\ L+2})+2J_{FA}\boldsymbol{S}_{2,1}^{\ L}+\boldsymbol{H}$
(F)	$\boldsymbol{S}_{1,2}^{\ j=L+2\to N-1}$	$2J_{FF}(2\boldsymbol{S}_{2,1}^{\ j}+\boldsymbol{S}_{2,1}^{\ j+1}+\boldsymbol{S}_{2,1}^{\ j-1})+\boldsymbol{H}$
(F)	$\boldsymbol{S}_{1,2}^{\ N}$	$2J_{FF}(2\boldsymbol{S}_{2,1}^{\ N}+\boldsymbol{S}_{2,1}^{\ N-1})+\boldsymbol{H}$

The equilibrium spin arrangement is given by $\partial \boldsymbol{S}/\partial t=0$. Using the simulation method given by the Landau–Lifshitz and Gilbert (LLG) equation, Koon determined the spin arrangement near the A/F interface as follows.

A dynamic equation for the spin $\boldsymbol{S}$ is

$$\partial \boldsymbol{S}/\partial t= -\gamma \boldsymbol{S}\times\{\boldsymbol{H}_{\text{eff}}-(\alpha/\gamma S)\ \partial \boldsymbol{S}/\partial t\}, \tag{2.21}$$

where γ is the gyromagnetic ratio and α is the damping constant. This equation can be written as follows:

$$\partial \boldsymbol{S}/\partial t= -\gamma\,[\boldsymbol{S}\times\boldsymbol{H}_{\text{eff}}-(\alpha/S)\boldsymbol{S}\times(\boldsymbol{S}\times\boldsymbol{H}_{\text{eff}})+(\alpha/S)^2\boldsymbol{S}\times(\boldsymbol{S}\times\partial \boldsymbol{S}/\partial t)]. \tag{2.22}$$

Further, this equation is modified as

$$(1+\alpha^2)\ \partial \boldsymbol{S}/\partial t= -\gamma \boldsymbol{S}\times[\boldsymbol{H}_{\text{eff}}+(\alpha/S)(\boldsymbol{S}\times\boldsymbol{H}_{\text{eff}})](\equiv \boldsymbol{T}). \tag{2.23}$$

Here $\boldsymbol{T}$ is the torque acting on spins, and the equilibrium condition of $\partial \boldsymbol{S}/\partial t=0$ is equivalent to T=0. Therefore, the equilibrium arrangement is determined as follows.

The initial S (t=0) is denoted as S_0 which is chosen appropriately. Starting with S_0, the initial H_{eff} is calculated from Table 2.1. Using S_0, $H_{eff}(t=0)$ and (2.23), the S's, H_{eff}'s and T at t (=Δt) are calculated. In general, using S's, H_{eff}'s and T at t_n (=$n\Delta t$), the increments $\Delta S_{1,2}^j$, $\Delta H_{eff\,1,2}^j$ and $\Delta T_{1,2}^j$ are calculated from Table 2.1 and by the following equations:

$$\Delta S_{1,2}^j = S_{1,2}^j(t_n+\Delta t) - S_{1,2}^j(t_n),$$
$$\Delta T_{1,2}^j = -\gamma \Delta S_{1,2}^j \times [H_{eff\,1,2}^j + \alpha(S_{1,2}^j \times H_{eff\,1,2}^j)]$$
$$-\gamma S_{1,2}^j \times [\Delta H_{eff\,1,2}^j + \alpha(\Delta S_{1,2}^j \times H_{eff\,1,2}^j + S_{1,2}^j \times \Delta H_{eff\,1,2}^j)],$$

Equation (2.23) is

$$(1+\alpha^2)\, \Delta S_{1,2}^j/\Delta t - \Delta T_{1,2}^j = T_{1,2}^j,$$

and this can be written with $\Delta S_{1,2}^j$, $\Delta H_{eff\,1,2}^j$ and $\Delta T_{1,2}^j$ as follows:

$$\begin{aligned}&\{(1+\alpha^2)/\Delta t\}\Delta S_{1,2}^j + \gamma\Delta[S_{1,2}^j \times H_{eff1,2}^j]\\ &+\gamma\alpha[\Delta(S_{1,2}^j \cdot H_{eff1,2}^j)S_{1,2}^j + (S_{1,2}^j \cdot H_{eff1,2}^j)\Delta S_{1,2}^j - \Delta H_{eff1,2}^j]\\ &= -\gamma S_{1,2}^j \times H_{eff1,2}^j - \gamma\alpha[(S_{1,2}^j \cdot H_{eff1,2}^j)S_{1,2}^j - H_{eff1,2}^j],\end{aligned} \tag{2.24}$$

where we assume $(S_{1,2}^j)^2=1$. Equation (2.24) is a set of linear simultaneous equations in $\Delta S_{1,2}^j$ and can be solved by the usual method with choosing suitable Δt, γ and α, which determine the rate of convergence. We repeat this procedure successively for T=0.

Koon's calculated results are introduced in the case of N, L=15. As boundary conditions he assumed the spins on the outside layers ($j\leq 1$) of the AL are directed along the easy axis of [0 0 1] and on the outside layers ($30\leq j$) of the FL are aligned uniformly along M (i.e., the direction of the induced anisotropy of the FL by the field-cooling treatment) and the angle between [0 0 1] and M is φ. Further he assumed $J_{FF}=-J_{AF}=-J_{AA}=1$ meV and K=0.05 meV(>0). The obtained total energy E is expressed as $E=c(\varphi-90°)^2$ with $c=1.98\times10^{-5}$ meV/deg^2. The energy minimum is at φ= 90° and thus the stable direction of M is perpendicular to the easy axis [0 0 1] of the AL. Namely, the easy axis of the ferromagnetic layers becomes perpendicular to the easy axis of the AL. In Fig. 2.7 the obtained spin arrangements on

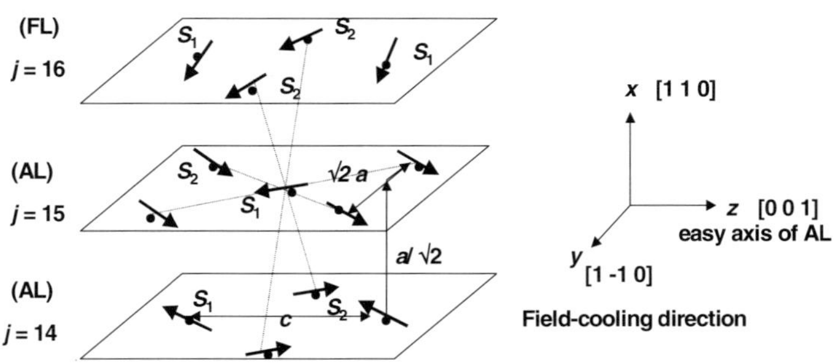

Fig. 2.7. Schematic presentation of spin arrangement on A and F layers near the A/F interface. The easy axis of ferromagnetic and of antiferromagnetic layers are perpendicular to each other. Spins swing around their easy axis, i.e., frustrated spin arrangement occurs.

the AL and the FL near the A/F interface are shown and so-called frustrated spin arrangements are created in the neighboring layers of j=15. For comparison he calculated spin arrangements on pure FLand pure AL, i.e., except for the interface all layers have homogeneous arrangements. In the pure case the total energy is $E=c\varphi^2$ with $c=2.10\times10^{-5}$ meV/deg^2. The frustrated structure shows a lower total minimum energy than that of pure layers by 0.12 meV/deg^2.

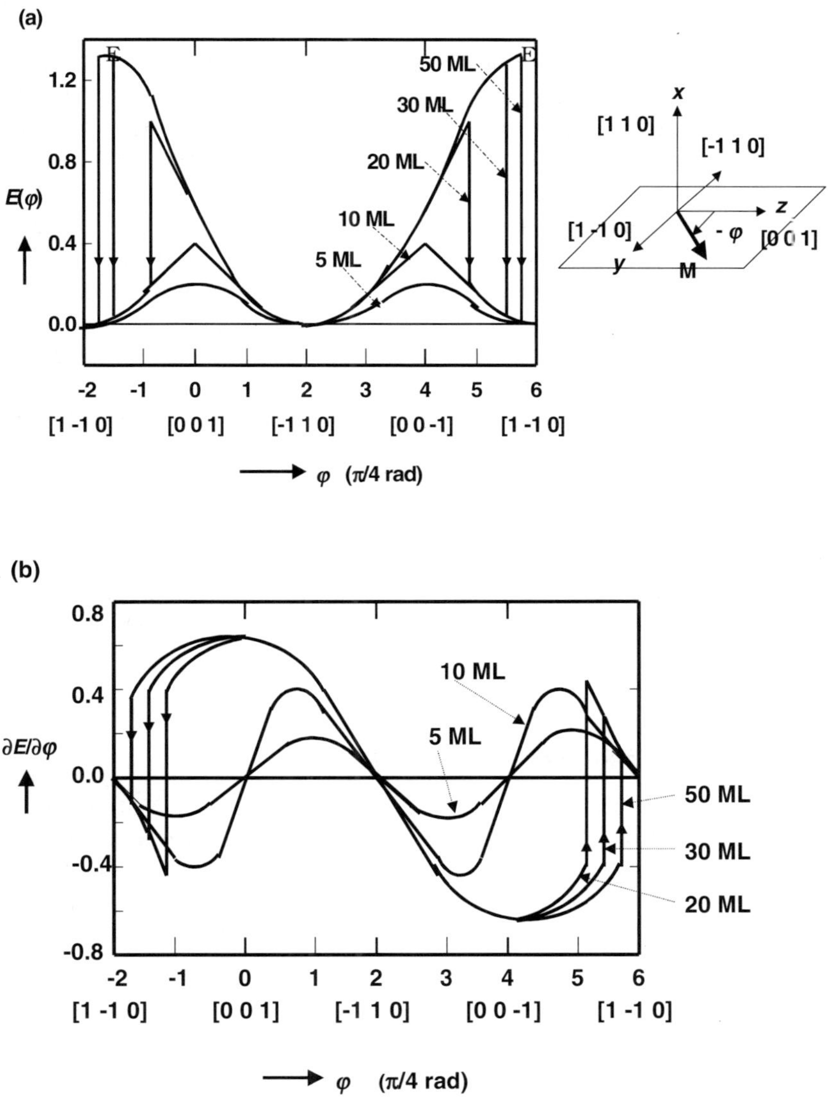

Fig. 2.8. (**a**) E(φ) curves and (**b**) Torque curves for various thickness L(in ML) of A-layers (Koon[2]).

Spin pinning of $\boldsymbol{M}$ by the AL is examined by assuming $J_{FF}=\infty$ (perfectly aligned spins on the FL), $J_{AF}=J_{AA}=-1$ meV, the tilt angle of $\boldsymbol{M}$ from [0 0 1] is φ, and the thickness of the AL is L (ML=multilayer). Since the stable directions of the $\boldsymbol{M}$ are $\varphi=\pm 90°$, putting $\varphi=90°$([-1 1 0] direction) initially, the stable spin arrangement of the AL and the total energy E are calculated. Then, shifting to $\varphi=90°+\Delta\varphi$ and starting from the above obtained arrangement as initial arrangement again the stable arrangement and total energy due to the shift $\Delta\varphi$ are calculated and thus the total energy is determined as functions of φ.

The calculated $E(\varphi)$ are shown in Fig. 2.8(a) for various thicknesses L in the range of 5–50 ML. If we start from the other stable position (i.e., $\varphi=-90°$([1 -1 0] direction)), the results are the same except the $E(\varphi)$ curves shift by -180°. The $E(\varphi)$ curves have minima at $\varphi= 90°$([-1 1 0]), and -90°(=270°[1-1 0]). In the case of $L\leq 10$ ML the $E(\varphi)$ curves are continuous. But in the case of $L>10$ ML the $E(\varphi)$ curves become discontinuous at angle φ_{cr}, and thus $\boldsymbol{M}$ is pinned at 90°in the range of $-\varphi_{cr}<\varphi<\varphi_{cr}$ and switches from 90°to another stable position (90°±180°) at the critical angle$\pm\varphi_{cr.}$ In the present case the estimated domain wall thickness is $w\cong(4J/K)^{1/2}=9$ ML and a thinner AL than the wall thickness can't pin $\boldsymbol{M}$ but a thicker AL pins $\boldsymbol{M}$ by its magnetic anisotropy. For example, the thick AL of 50 ML pins $\boldsymbol{M}$ in the range of 90°±172°, namely, almost perfectly. Fig. 2.8(b) shows the torque curve of $\boldsymbol{M}$ received from the AL which is another representation of Fig. 2.8(a). In practice field-cooling treatment causes antiferromagnetic domain patterns in the AL, in which the easy axis is perpendicular to the field-cooled direction. So Koon's results seem to be natural but in order to compare experiments on epitaxial layers with this theory the field-cooling treatment has to be done along [-1 1 0] or [1 -1 0]. In the case of polycrystals the field-cooling treatment causes an induced anisotropy due to domains in the AL, in which the easy axis [0 0 1] of each domain is directed nearly perpendicular to the direction of the field-cooling treatment.

2.1.2.2 Uncompensated Case: (1 0 0)-Interface Plane

The spins on the interface layers of the (1 0 0) plane are in ferromagnetic arrangement shown in Fig. 2.9. Each layer is uncompensated and has a spacing distance of $a/2$. The number of spins per unit area is $1/ac$ and eight of the nearest neighbor spins are in adjacent layers. If we assume L and N are even numbers to simplify expressions, the total magnetic energy is

$$\begin{aligned} E = &-4J_{AA}[\boldsymbol{S}_1^{\,1}\cdot\boldsymbol{S}_2^{\,2}+K(S_{1z}^{\,j})^2]-4J_{AA}\sum_{j=2}^{L-1}[\boldsymbol{S}_1^{\,j}\cdot(\boldsymbol{S}_2^{\,j+1}+\boldsymbol{S}_2^{\,j-1})+K(S_{1z}^{\,j})^2] \\ &-4J_{AA}\boldsymbol{S}_1^{\,L}\cdot\boldsymbol{S}_2^{\,L-1}-4J_{AF}\boldsymbol{S}_1^{\,L}\cdot\boldsymbol{S}_2^{\,L+1}-K(S_{1z}^{\,L})^2-4J_{FA}\boldsymbol{S}_2^{\,L+1}\cdot\boldsymbol{S}_1^{\,L}-4J_{FF}\,\boldsymbol{S}_2^{\,L+1}\cdot\boldsymbol{S}_1^{\,L+2} \\ &-4J_{FF}\sum_{L+2}^{L+N-1}[\boldsymbol{S}_1^{\,j}\cdot(\boldsymbol{S}_2^{\,j+1}+\boldsymbol{S}_2^{\,j-1})]-4J_{FF}\boldsymbol{S}_2^{\,N}\cdot\boldsymbol{S}_1^{\,N-1}-\sum_{j=1}^{L+N}\boldsymbol{H}\cdot\boldsymbol{S}_1^{\,j}. \end{aligned} \tag{2.25}$$

In cases of odd L and N the expression is complex but essentially similar.

The effective fields $\boldsymbol{H}_{\text{eff }1,2}^{\,j}$ are tabulated in Table 2.2. Further discussion is abridged, but because the nearest neighbor spins are in adjacent layers, the exchange interaction couples spins on adjacent layers. Thus, the uncompensated AL couples with the FL by J_{AF} and spins on the FL are pinned along the easy axis of the AL.

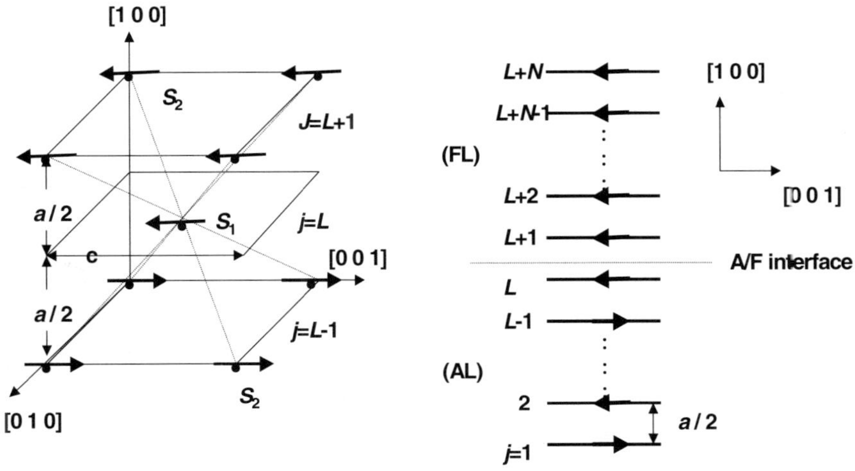

(a) (b)

Fig. 2.9. (1 0 0)-layer structure near an A/F interface

Table 2.2. Effective magnetic fields for $\boldsymbol{S}_{1,2}{}^{j}$ on the (0 0 1)plane.

Layer	$\boldsymbol{S}_{1,2}{}^{j}$	$\boldsymbol{H}_{\text{eff}\,1,2}{}^{j}$
(A)	$\boldsymbol{S}_{1,2}{}^{j=1\rightarrow L-1}$	$4J_{AA}\,(\boldsymbol{S}_{2,1}{}^{j+1}+\boldsymbol{S}_{2,1}{}^{j-1})+2K\boldsymbol{S}_{1,2z}{}^{j}+\boldsymbol{H}$
(A)	$\boldsymbol{S}_{1,2}{}^{L}$	$4J_{AA}\,\boldsymbol{S}_{2,1}{}^{L-1}+4J_{AF}\,\boldsymbol{S}_{2,1}{}^{L+1}+2K\boldsymbol{S}_{1,2z}{}^{L}+\boldsymbol{H}$
(F)	$\boldsymbol{S}_{1,2}{}^{L+1}$	$4J_{FF}\,\boldsymbol{S}_{2,1}{}^{L+2}+4J_{FA}\,\boldsymbol{S}_{2,1}{}^{L}+\boldsymbol{H}$
(F)	$\boldsymbol{S}_{1,2}{}^{j=L+2\rightarrow N-1}$	$2J_{FF}\,(\boldsymbol{S}_{2,1}{}^{j+1}+\boldsymbol{S}_{2,1}{}^{j-1})+\boldsymbol{H}$
(F)	$\boldsymbol{S}_{1,2}{}^{N}$	$2J_{FF}\,\boldsymbol{S}_{2,1}{}^{L-1}+\boldsymbol{H}$

2.2 Free-Electron Model of the GMR Effect

The basic structure of GMR devices is a nonmagnetic metal film sandwiched between two ferromagnetic metal films. The electric conduction of this structure is described by the Boltzmann transport equation of electrons with spin-dependent relaxation time $\tau^{\uparrow,\downarrow}$, where $\uparrow$ and $\downarrow$ indicate majority and minority electrons in ferromagnetic metals. The conductivity of this structure depends on the angle 2ϕ between the magnetizations of both ferromagnetic films. If an applied field changes this angle 2ϕ, the conductivity is also changed with applied field. This is the magnetoresistance effect, and its ratio is about 10% or more, and therefore it is called giant magnetoresistance (GMR) in comparison with AMR.

In this section the physics of GMR is explained. In Sect. 2.2.1 conductivity expressions for metals and metal thin films are described, and in Sect. 2.2.2 spin-dependent conductivity in ferromagnetic metals (two-current model) is disscussed. After a short review of the MR effect in metal-multilayers the free electron model of the GMR effect is given in Sect. 2.2.3. This topic is treated in a recent publication [3], in which the advanced physics of GMR is describe in detail.

2.2.1 Conductivity of Metals and Metal Films

Bulk Metal

Electric conduction in bulk metals is described briefly in this section. In the free-electron approximation the distribution function $f(\boldsymbol{r},\boldsymbol{v};t)$ of conduction electrons with mass m, charge $-e$ and velocity $\boldsymbol{v}$ is expressed by the following Boltzmann transport equation:

$$\partial f/\partial t + \boldsymbol{v}\cdot\nabla f - e\boldsymbol{E}\cdot\partial f/\partial \boldsymbol{p} = -(f - f_0)/\tau, \tag{2.26}$$

where $f_0(\boldsymbol{v})$ is the distribution function of the homogeneous steady state with no applied field and is given by the Fermi–Dirac function: $f_0 = 1/[\exp(\varepsilon-\varepsilon_F)/kT+1]$ with $\varepsilon = m\,\boldsymbol{v}^2/2$ and the Fermi energy ε_F. In an electric field $\boldsymbol{E}$ (x-axis) the distribution function f shifts to $f(=f_0+g)$ and from (2.26) the shift function $g(\boldsymbol{v})$ is given by

$$g(\boldsymbol{v}) = (eE\tau/m)(\partial f_0/\partial v_x), \tag{2.27}$$

where τ is the relaxation time of electrons scattered by impurities or thermal phonons in metals.

The numbers of electrons, which have velocity $\boldsymbol{v}$ within the infinitesimal space $d\boldsymbol{v}$, is expressed by using the state density $D(\boldsymbol{v})$ as $dn=D(\boldsymbol{v})d\boldsymbol{v}=[2(m/\mathrm{h})^3]f d\boldsymbol{v}$ with Planck's constant h. Therefore, the electric current caused by $\boldsymbol{E}$ is

$$J = -e[2(m/\mathrm{h})^3]\int d\boldsymbol{v}\; v_x g(\boldsymbol{v}). \tag{2.28}$$

Using the polar coordinates (v,θ,ϕ) in velocity space, J is expressed as follows,

$$\begin{aligned} J &= -(e^2E\tau/m)[2(m/\mathrm{h})^3] \\ &\quad\times\int^\infty_0 dv\, v^3\{\partial f_0(v)/\partial v\}\int^{2\pi}_0 d\phi\cos^2\phi\int^\pi_0 d\theta\sin^3\theta \\ &= (4\pi/3)(e^2E\tau/m)v_F^3[2(m/\mathrm{h})^3] = e^2E\tau n/m = \sigma_0 E \quad \text{(Ohm's law)} \end{aligned} \tag{2.29}$$

Where $\sigma_0 \equiv e^2\tau n/m$ (conductivity), $v_F=(\mathrm{h}/2\pi m)(3\pi^2 n)^{1/3}$ (Fermi velocity) and $\varepsilon_F = (mv_F^2)/2 = [(\mathrm{h}/2\pi)^2/2m](3\pi^2 n)^{2/3}$ (Fermi energy), respectively. Because $\partial f_0/\partial\varepsilon \cong -\delta(\varepsilon-\varepsilon_F)$ the current J is carried by electrons on the Fermi surface and σ_0 is determined by τ or mean free path $\lambda=v_F\tau$. The density n of conduction electrons in the metal is about 10^{22}.

Metal Thin Film

Fuchs and Sondheimer [4] studied the electric conductivity of a metal thin film with thickness t and an electric field $\boldsymbol{E}$ in the film plane. Choosing the x-axis and z-axis parallel to $\boldsymbol{E}$ and the normal of the film plane, respectively, the shift function $g(z,\mathbf{v})$ becomes

$$v_z \partial g(z,\mathbf{v})/\partial z + g/\tau = (eE/m)\, \partial f_0/\partial v_x. \tag{2.30}$$

The general solution of (2.30) is given by

$$g(z,\mathbf{v}) = (eE\tau/m)\{\partial f_0(v)/\partial v_x\}[1+F(\mathbf{v})\exp(-z/v_z\tau)], \tag{2.31}$$

where $F(\mathbf{v})$ is an integral constant determined by the boundary conditions.

Discriminating the shift functions for positive v_z and for negative v_z by $g(z,\mathbf{v})_+$ and $g(z,\mathbf{v})_-$, (2.31) is

$$g(z,\mathbf{v})_\pm = (eE\tau/m)\{\partial f_0(v)/\partial v_x\}[1+F(\mathbf{v})_\pm\exp\{-z/(\pm v_z\tau)\}]. \tag{2.32}$$

At both surfaces of the thin film, electrons are partly reflected specularly or scattered diffusively. Therefore, the boundary conditions are given with specular factor p as follows,

$$pg(0,\mathbf{v})_- = g(0,\mathbf{v})_+ \quad \text{at } x=0, \quad pg(t,\mathbf{v})_+ = g(t,\mathbf{v})_- \quad \text{at } x=t. \tag{2.33}$$

These conditions are

$$p[1+F(\mathbf{v})_-] = [1+F(\mathbf{v})_+],$$

$$p[1+F(\mathbf{v})_+\exp\{-t/(v_z\tau)\}] = [1+F(\mathbf{v})_-\exp\{t/(v_z\tau)\}]. \tag{2.34}$$

Therefore, integral constants $F(\mathbf{v})_\pm$ are determined as

$$F(\mathbf{v})_+ = -(1-p)/[1-p\exp\{-t/(v_z\tau)\}],$$

$$F(\mathbf{v})_- = F(\mathbf{v})_+\exp\{-t/(v_z\tau)\}. \tag{2.35}$$

If there is specular reflection then p=1, $F(\mathbf{v})_\pm$ are zero, i.e., electrons behave just as if there were no surfaces and the conductivity is the same as that of the bulk metal mentioned previously.

The current distribution $J(z)$ along the z-axis is

$$J(z) = -e[2(m/h)^3]\int \mathbf{dv}\, v_x g(z,\mathbf{v}). \tag{2.36}$$

Using the polar coordinates (v,θ,ϕ), $J(z)$ is given by

$$\begin{aligned} J(z) = &-e[4(m/h)^3](eE\tau/m)\int^\infty_0 dv\, v^3\{\partial f_0(v)/\partial v\}\int^{2\pi}_0 d\phi \cos^2\phi \\ &\times\int^{\pi/2}_0 d\theta \sin^3\theta[1-(1-p)\exp\{-t/2(\tau v\cos\theta)\} \\ &\times\cosh\{(z-t/2)/(\tau v\cos\theta)\}/[1-p\exp\{-t/(\tau v\cos\theta)\}]] \\ = &E\sigma_0[1-(3/2)\int^{\pi/2}_0 d\theta\sin^3\theta[1-(1-p)\exp\{-t/2(\lambda\cos\theta)\} \\ &\times\cosh\{(z-t/2)/(\lambda\cos\theta)\}/[1-p\exp\{-t/(\lambda\cos\theta)\}]], \end{aligned} \tag{2.37}$$

where $v_z = v\cos\theta$. Thus the conductivity σ of thin films is

$$\begin{aligned} \sigma = \int^t_0 dz\, J(z)/Et = \sigma_0[1-(3\lambda/2t)\int^{\pi/2}_0 d\theta\sin^3\theta\cos\theta \\ \times(1-p)[1-\exp\{-t/(\lambda\cos\theta)\}]/[1-p\exp\{-t/(\lambda\cos\theta)\}]]. \end{aligned} \tag{2.38}$$

The conductivity ratio of σ/σ_0 is determined as a function of the ratio q $(=t/\tau)$ and p. The calculated results of Sondheimer *et al.* are shown in Fig. 2.10, which shows σ_0/σ as functions of q and p.

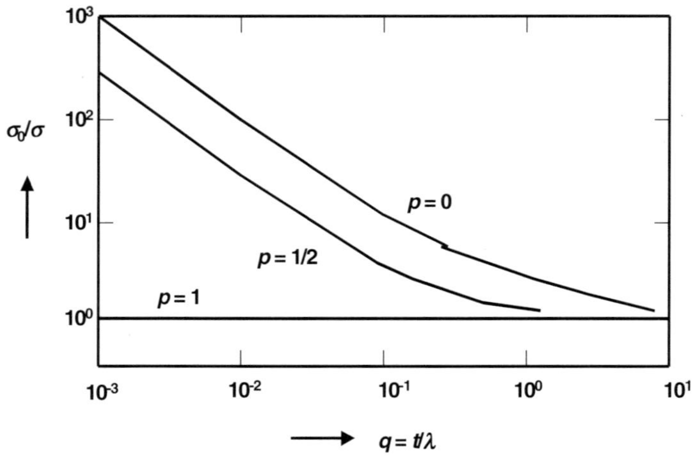

Fig. 2.30. The σ_0/σ as functions of $q(=t/\tau)$ and p

In extreme cases (2.38) reduces to

$$q >> 1 \qquad \sigma_0/\sigma = 1 + 3(1 - p)/8q, \tag{2.39}$$

and

$$q << 1 \quad \sigma_0/\sigma = \{4(1 - p)/3(1 + p)\}/q \log(1/q). \tag{2.40}$$

In case of $q>>1$ (thick film) $\sigma_0/\sigma \cong 1$ and second term on the right in (2.39) shows the surface effect, but for $q<<1$ (thin film) $\sigma_0/\sigma \rightarrow \{4(1\text{-}p)/3(1+p)\}\lambda/t$ and is determined mainly by the surface scattering p of the film.

2.2.2 Conductivity of a Ferromagnetic Metal (Two-Current Model)

In nonmagnetic metals conduction electrons are scattered by imperfections such as phonons, impurities and lattice defects and the resistivity is expressed as the additive law $\rho(T) = \rho_0 + \rho_p(T)$ (Matthiessen's law). In this relation the temperature-independent term ρ_0 is called the residual resistivity, which is due to scattering by impurities and lattice defects, and the second term $\rho_p(T)$ is temperature dependent and due to scattering by thermal phonons.

In ferromagnetic metals conduction electrons are scattered further by magnons and by magnetic impurities originate in interactions between magnetic moments of these scattering centers and spins of conduction electrons. Electrons with parallel spin to the total magnetization of ferromagnetic metals, namely, electrons with majority spin indicated by ↑ and electrons with minority spin by ↓. At sufficiently lower temperature than T_c magnon scattering becomes negligibly small and the majority spin ↑ and minority spin ↓ electrons don't mix with each other in scattering processes. Therefore, an electric current is carried by electrons with ↑-spin and with ↓-spin, independently, and thus the conductivity is expressed as

$$\sigma = \sigma^{\uparrow} + \sigma^{\downarrow}, \tag{2.41}$$

where $\sigma^{\uparrow,\downarrow} = e^2\tau^{\uparrow,\downarrow}n^{\uparrow,\downarrow}/m*$ with $m*$ an effective mass (the two-current model of Mott [5]).

The electric conductivity in ferromagnetic metals with various magnetic impurities is considered within the two-current model. In a ferromagnetic metal the majority spin $d^{\uparrow}$-band and the minority spin $d^{\downarrow}$-band shift with respect to each other due to exchange interaction between d-electrons. The $d^{\uparrow}$-band shifts down to the $d^{\downarrow}$-band by the exchange energy ε_x and the Fermi energy ε_F locates above the $d^{\uparrow}$-band (completely filled band) and in the $d^{\downarrow}$-band (partially filled band) as shown schematically in Fig. 2.11. The conduction electrons of the $s^{\downarrow}$-band are scattered by magnetic impurities to the $d^{\downarrow}$-band, where electrons have a large effective mass m* and contribute slightly to the electric conduction, while electrons of the $s^{\uparrow}$-band don't scatter to the $d^{\uparrow}$-band due to the completely filled $d^{\uparrow}$-band at ε_F.

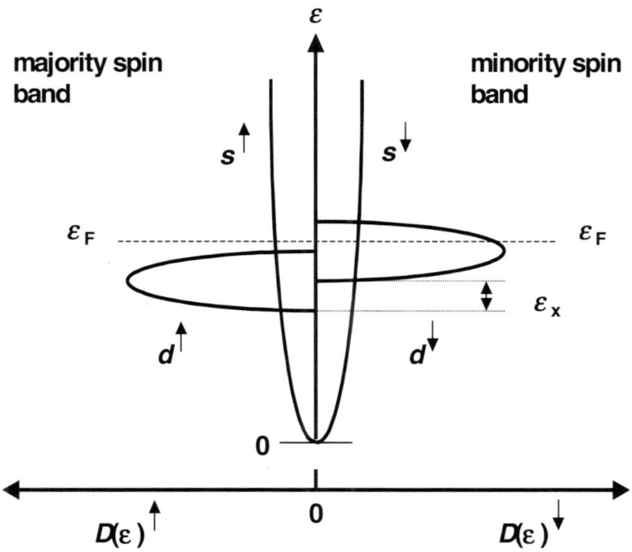

Fig. 2.11. Schematic presentation of spin-polarized bands in a ferromagnetic metal

Therefore, $\tau^{\downarrow} < \tau^{\uparrow}$ and thus conductance asymmetry occurs and the asymmetry factor is $\alpha = \sigma^{\uparrow}/\sigma^{\downarrow} (= \rho^{\downarrow}/\rho^{\uparrow}) > 1$.

Calculated energy bands for Fe, Co, Ni and Cu are shown in Fig. 2.12. The $d^{\uparrow}$-band of Co and Ni locate below ε_F, therefore conductivities in Ni and Co show large asymmetry. But for Fe the majority $d^{\uparrow}$-band is incompletely filled, thus the difference between the state densities for the ↑- and ↓- bands at ε_F is small and the asymmetry factor of Fe is also small. Experimental data on the asymmetry factor α are tabulated for various dilute impurities in the host metal of (a) Ni, (b) Co and (c) Fe in Table 2.3. These data are determined from measurements of residual resistances $\rho_0^{\uparrow}$ and $\rho_0^{\downarrow}$, and by $\alpha = \rho_0^{\downarrow}/\rho_0^{\uparrow}$ [6].

At high temperature, spins flop between the ↑- spin state and the ↓- spin state by magnon scattering, and the two-current model doesn't hold due to spin mixing. In the case Campbell and Fert [6] show that the resistivity is given by

$$\rho = [\rho^{\uparrow}\rho^{\downarrow}+\rho^{\uparrow\downarrow}(\rho^{\uparrow}+\rho^{\downarrow})]/(\rho^{\uparrow}+\rho^{\downarrow}+4\rho^{\uparrow\downarrow})$$
$$\cong [\rho^{\uparrow}\rho^{\downarrow}/(\rho^{\uparrow}+\rho^{\downarrow})]+[(\alpha-1)/(\alpha+1)]^2\rho^{\uparrow\downarrow}, \tag{2.42}$$

where $\rho^{\uparrow\downarrow}$ is caused by spin-flip scattering.

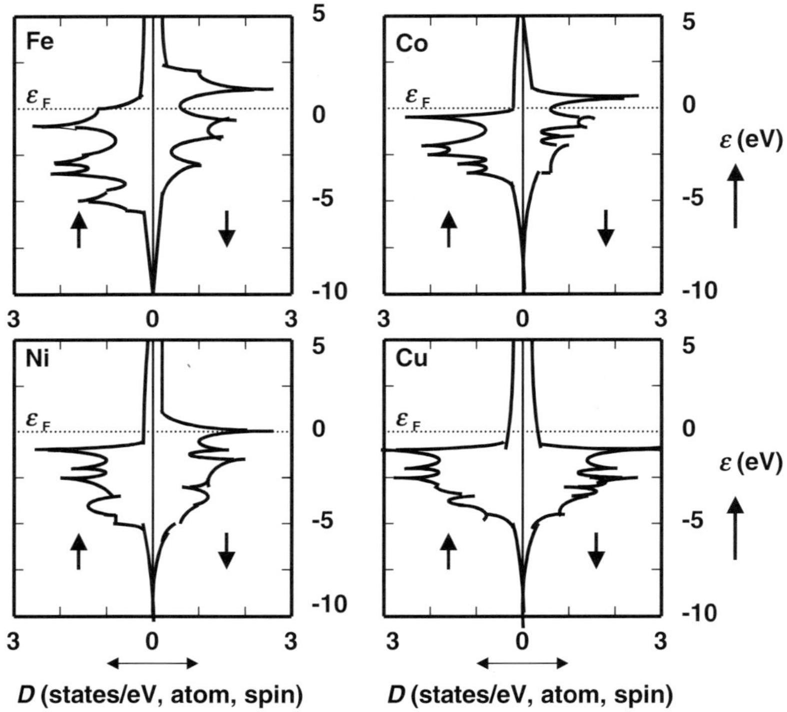

Fig. 2.12. Calculated spin-polarized bands of 3d-transition metals; Fe, Co, Ni, Cu

Table 2.3. Experimental data for the asymmetry factor α for various impurities in the host metal of (a) Ni, (b) Co, (c) Fe

Element	Ti	V	Cr	Mn	Fe	Co	Ni	Cu
(a) Ni-host	0.9	0.45	0.21	6.3	11	13	*	2.9
(b) Co-host	1.4	1.0	0.3	0.8	12	*	–	–
(c) Fe-host	0.25	0.12	0.17	0.09	*	1	3	–

2.2.3 Two-Current Model for the GMR Effect

There are two kinds of magnetoresistance effect in ferromanetic metals. One is the anisotropic magnetoresistance (AMR) effect in a ferromagnetic metal monolayer with uniaxial anisotropy and the other is the giant magnetoresistance (GMR) in a metal multilayer. Brief descriptions of the AMR and GMR effect are given in Sects. 1.1.1 and 1.1.2.

In GMR devices the sensing current flows in layer planes (CIP) or flows perpendicular to the layer planes (CPP). Usually, the MR ratio of the CPP mode is larger than that of the CIP mode, but the GMR effect of the CPP mode is too difficult to use in practical devices because of the extremely small resistance of such devices. In this section the GMR effect in the CIP mode is considered on the basis of the two-current model. A first simple explanation for GMR in a metal multilayer is given in Sect. 2.2.3.1 and then the quantitative model for the GMR ratio in a spin-valve structure is explained in Sect. 2.2.3.2. This model can be extended to various type of multilayers.

2.2.3.1 GMR Effect in a Metal Multilayer

The resistances of a ferromagnetic/nonmagnetic (F/C)-metal multilayer are denoted by $R_{\uparrow\uparrow}$ and $R_{\uparrow\downarrow}$ corresponding to parallel and to antiparallel alignment of the magnetizations of successive layers, respectively. The electron trajectories for an up spin (↑) electron and for a down spin (↓) electron are shown schematcally in Fig. 2.13. If the electron spin differs from the magnetization of the passing layer,

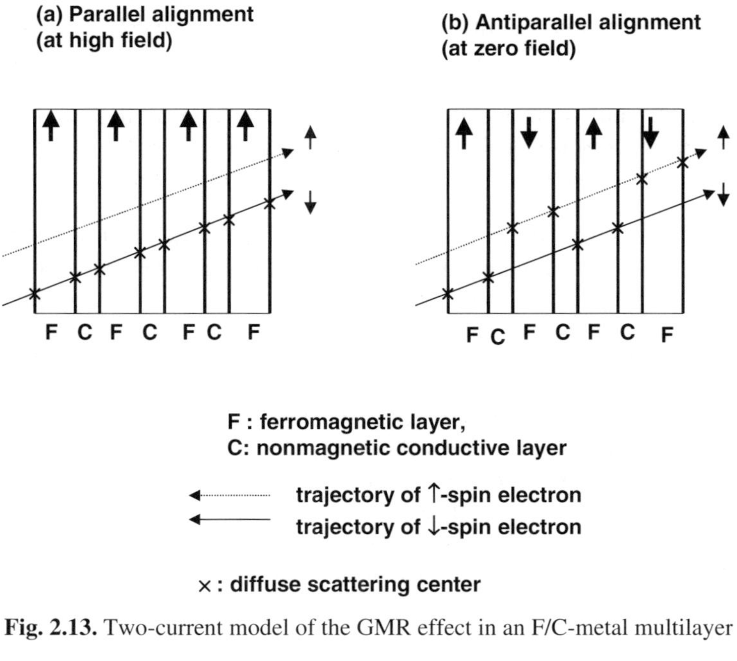

Fig. 2.13. Two-current model of the GMR effect in an F/C-metal multilayer

conduction electrons are scattered more strongly. We denote by $R^{\uparrow}$ and $R^{\downarrow}$ resistance of majority spin electrons and of minority spin electrons in each magnetic layer, $R^{\uparrow}<R^{\downarrow}$; and if the mean free path λ of the conduction electrons is larger than the period of the F/C multilayer, $R_{\uparrow\uparrow} = (1/R^{\uparrow}+1/R^{\downarrow})^{-1}$ and $R_{\uparrow\downarrow} =(R^{\uparrow}+ R^{\downarrow})/4$. In this case magnetoresistance ratio is given by

$$\chi = (R_{\uparrow\downarrow} - R_{\uparrow\uparrow})/R_{\uparrow\uparrow} = (R^{\downarrow}-R^{\uparrow})^2/4R^{\downarrow}R^{\uparrow} = (1-\alpha)^2/4\alpha, \tag{2.43}$$

where $\alpha=R^{\downarrow}/R^{\uparrow}$. χ is always positive and $R_{\uparrow\downarrow} > R_{\uparrow\uparrow}$. If there are spin-independent resistances R_0, the MR ratio becomes small.

Scattering of conduction electrons in ferromagnetic metals is discussed in Sect. 2.2.2, but in an F/C-metal multilayer the effective scattering for χ includes the diffuse scattering at interfaces of multilayers. Various studies of interface scattering have been carried out considering electronic and/or structural matching at interfaces. In the following section the details of interface scattering are omitted but scattering is considered assuming the coefficients of spin-dependent diffuse scattering $D^{\uparrow}$ and $D^{\downarrow}$.

2.2.3.2 GMR Effect in a Spin-Valve Structure

The basic structural unit of GMR devices consists of a nonmagnetic conductive layer sandwiched between two ferromagnetic metal layers. The current in this unit is dependent on the angle 2ϕ between the magnetizations of both ferromagnetic layers. The resistance of this unit is expressed by

$$R = R_{\uparrow\uparrow} + \Delta R\,(1-\cos 2\phi)/2, \tag{2.44}$$

where $\Delta R=R_{\uparrow\downarrow} - R_{\uparrow\uparrow}$, and $R_{\uparrow\downarrow}$ and $R_{\uparrow\uparrow}$ are the resistances for antiparallel ($2\phi=\pi$) and parallel ($2\phi=0$) magnetizations of the both magnetic layers, respectively. The magnetoresistance ratio $\alpha(=\Delta R/R_{\uparrow\uparrow})$ is larger than 10% and is called giant magnetoresistance (GMR) as compared to that of the AMR. The GMR effect depends only on the angle 2ϕ and is independent the current direction as indicated in (2.44). The quantitative discussion of the GMR effect is given on the basis of the Huchs–Sondheimer treatment of electric conduction in metal thin films (see Sect. 2.2.1).

A typical GMR device is a spin valve with the structure shown in Fig. 2.14. The structure is basically a ferromagnetic layer (PL) with pinned magnetization, another ferromagnetic layer (FL) with free magnetization and a nonmagnetic metal spacer layer (SL) for separating magnetically both ferromagnetic layers. In an applied magnetic field $\boldsymbol{H}$ the magnetization $\boldsymbol{M}_{SP}$ of the PL is pinned by unidirectional anisotropy due to an adjacent antiferromagnetic layer AL but the magnetization $\boldsymbol{M}_{SF}$ of the FL rotates freely following $\boldsymbol{H}$. Thus, the resistance of a spin-valve device changes with following $\boldsymbol{H}$. In Fig. 2.14, $\boldsymbol{M}_{SP}$ is pinned along the perpendicular direction to the current $\boldsymbol{J}$ (x-axis)and $\boldsymbol{M}_{SF}$ rotates by an applied field $\boldsymbol{H}$ and thus the angle 2ϕ between both magnetizations changes with following $\boldsymbol{H}$. Barnas *et al.* [7] considered this MR effect in the basic trilayer structure on the basis of the free electron model with spin-dependent relaxation time and by assuming the quantization axes of PL and FL change their directions by 2ϕ at the center of the SL.

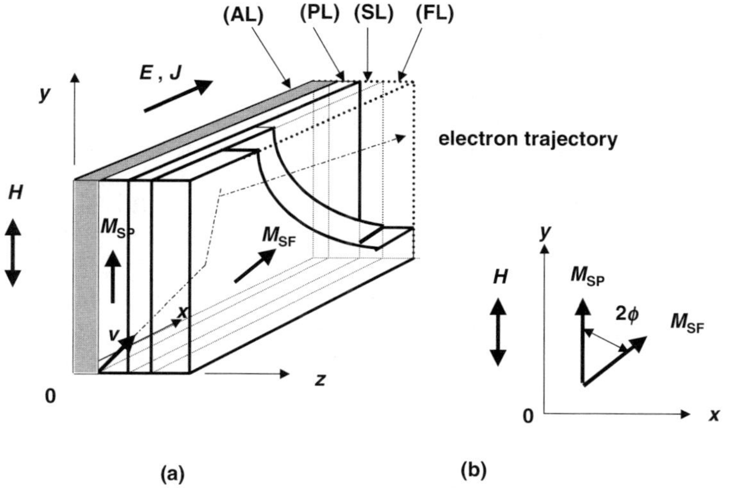

Fig. 2.14. Spin-valve structure: (**a**) schematic presentation of a spin valve, in which (AL) is antiferromagnetic layer, (PL) a pinned magnetization $\boldsymbol{M}_{SP}$ layer, (SL) a nonmagnetic spacer layer and (FL) a free magnetization $\boldsymbol{M}_{SF}$ layer, (**b**) a configuration of the magnetizations in an applied field $\boldsymbol{H}$

Calculation Model and Electric Conductivity

In rectangular coordinates we take the normal to the layer as the z-axis, applied electric field $\boldsymbol{E}$ as the x-axis and the applied magnetic field $\boldsymbol{H}$ as the y-axis. In the

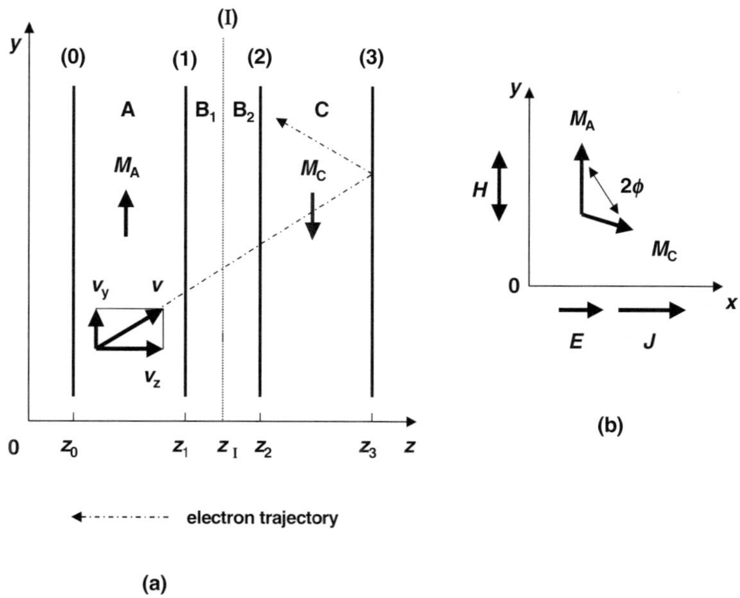

Fig. 2.15. Basic calculation model of a spin valve: (**a**) layer structure and (**b**) magnetization configuration

ith layer electrons have Fermi velocity $\boldsymbol{v}_{Fi}$ and relaxation time τ_i and electrons in the ferromagnetic layers have the spin-dependent relaxation time $\tau_i^{\uparrow,\downarrow}$, i.e., the mean free path $\lambda_i^{\uparrow,\downarrow}$ $(=v_{Fi}\,\tau_i^{\uparrow,\downarrow})$. We assume the quantization axes change their direction by 2ϕ on a hypothetical plane at the center of the spacer layer SL. This calculation model is shown in Fig. 2.15, and its surfaces and interfaces are numbered by i (=0,1,2,3) and layers by A,B,C. The z-positions of each surfaces and interfaces are indicated as z_i. Layers A and C are ferromagnetic and B is nonmagnetic metal spacer layer and the magnetizations of A and C are denoted as $\boldsymbol{M}_A$ and $\boldsymbol{M}_C$.

In an electric field $\boldsymbol{E}$ the distribution function of electrons depends on the z-coordinate and the electron velocity $\boldsymbol{v}$ and is expressed as follows:

$$g(z,\boldsymbol{v})_\pm^{\uparrow,\downarrow} = (eE\tau^{\uparrow,\downarrow}/\mathrm{m})\{\partial f_0(\boldsymbol{v})/\partial v_x\}[1+F(\boldsymbol{v})_\pm^{\uparrow,\downarrow}\exp(-z/\pm v_z\tau^{\uparrow,\downarrow})]. \tag{2.45}$$

The current distribution $J(z)$ is

$$\begin{aligned} J(z) &= -e[2(m/\mathrm{h})^3][\Sigma_\pm^{\uparrow,\downarrow}\{\int \mathrm{d}\boldsymbol{v}\; g(z,\boldsymbol{v})_\pm^{\uparrow,\downarrow}\, v_x\}] \\ &= -e[2(m/\mathrm{h})^3][\,\Sigma_\pm^{\uparrow,\downarrow}\,(eE\tau^{\uparrow,\downarrow}/m)\int \mathrm{d}\boldsymbol{v}\;\{\partial f_0/\partial v_x\} \\ &\quad +\Sigma_\pm^{\uparrow,\downarrow}\,(eE\tau^{\uparrow,\downarrow}/m)\int \mathrm{d}\boldsymbol{v}\;\{\partial f_0/\partial v_x\}F(\boldsymbol{v})_\pm^{\uparrow,\downarrow}\, v_x \exp[-z/(\pm\tau^{\uparrow,\downarrow}\, v_z)]\}. \end{aligned} \tag{2.46}$$

Using polar coordinates this equation becomes

$$\begin{aligned} J(z) &= E[\Sigma^{\uparrow,\downarrow}\,(e^2\tau^{\uparrow,\downarrow}n/m)+\Sigma_\pm^{\uparrow,\downarrow}\,(\mathrm{e}\tau^{\uparrow,\downarrow}n/m) \\ &\quad \times(3/4)\int^\pi_0 \mathrm{d}\theta\, F(v_F)_\pm^{\uparrow,\downarrow}\sin^3\theta\exp(-z/\pm\lambda^{\uparrow,\downarrow}\cos\theta)] \\ &= E\Sigma^{\uparrow,\downarrow}\sigma^{\uparrow,\downarrow}\times[1+\Sigma_\pm(3/4)\int^\pi_0 \mathrm{d}\theta\, F(v_F)_\pm^{\uparrow,\downarrow}\sin^3\theta\exp(-z/\pm\lambda^{\uparrow,\downarrow}\cos\theta)], \end{aligned} \tag{2.47}$$

where $\tau v_z=\lambda\cos\theta$. The total current per unit length of y-axis is given by $J=\int J(z)\,\mathrm{d}z$, and the effective conductivity is

$$\begin{aligned} \sigma_{\mathrm{eff}} &= J/Et = (1/t)\,\Sigma_i\{\sigma_i t_i + (3/4)\,\Sigma_\pm^{\uparrow,\downarrow}\,\sigma^{\uparrow,\downarrow}{}_i\lambda^{\uparrow,\downarrow}{}_i\int^\pi_0 F(v_F)_\pm^{\uparrow,\downarrow} \\ &\quad \times[\exp(-z_{i-1}/\pm\lambda^{\uparrow,\downarrow}{}_{i-1}\cos\theta)-\exp(-z_i/\pm\lambda^{\uparrow,\downarrow}{}_i\cos\theta)]\sin^3\theta\cos\theta\,\mathrm{d}\theta\}. \end{aligned} \tag{2.48}$$

In the case of $\lambda_i \ll t_i$, interface scattering of the i-th layer contributes small to the resistance, therefore this equation becomes $\sigma_{\mathrm{eff}}\; t = \Sigma_i\sigma_i t_i$. Putting $z_{iA}^{\uparrow,\downarrow} \equiv \exp(z_i/\lambda_A^{\uparrow,\downarrow}\cos\theta)$, this conductivity is expressed as

$$\begin{aligned} \sigma_{\mathrm{eff}}\, t &= [(\sigma_A^{\uparrow}+\sigma_A^{\downarrow})t_A+\sigma_B t_B+(\sigma_C^{\uparrow}+\sigma_C^{\downarrow})t_C] \\ &\quad +(3/4)\{\int^{\pi/2}_0 \mathrm{d}\theta\sin^3\theta\cos\theta \\ &\quad \times\{[\sigma_A^{\uparrow}\lambda_A^{\uparrow}[F_{A+}^{\uparrow}(1/z_{0A}^{\uparrow}-1/z_{1A}^{\uparrow})+F_{A-}^{\uparrow}(z_{1A}^{\uparrow}-z_{0A}^{\uparrow})] \\ &\quad +\sigma_A^{\downarrow}\lambda_A^{\downarrow}[F_{A+}^{\downarrow}(1/z_{0A}^{\downarrow}-1/z_{1A}^{\downarrow})+F_{A-}^{\downarrow}(z_{1A}^{\downarrow}-z_{0A}^{\downarrow})] \\ &\quad +\sigma_B\lambda_B\{[(F_{B1+}^{\uparrow}+F_{B1+}^{\downarrow})(1/z_{1B1}-1/z_{IB1})+(F_{B1-}^{\uparrow}+F_{B1-}^{\downarrow})(z_{IB1}-z_{1B1})] \\ &\quad +[(F_{B2+}^{\uparrow}+F_{B2+}^{\downarrow})(1/z_{IB2}-1/z_{2B2})+(F_{B2-}^{\uparrow}+F_{B2-}^{\downarrow})(z_{2B2}-z_{IB2})]\} \\ &\quad +\sigma_C^{\uparrow}\lambda_C^{\uparrow}[F_{C+}^{\uparrow}(1/z_{2C}^{\uparrow}-1/z_{3C}^{\uparrow})+F_{C-}^{\uparrow}(z_{3C}^{\uparrow}-z_{2C}^{\uparrow})] \\ &\quad +\sigma_C^{\downarrow}\lambda_C^{\downarrow}[F_{C+}^{\downarrow}(1/z_{2C}^{\downarrow}-1/z_{3C}^{\downarrow})+F_{C-}^{\downarrow}(z_{3C}^{\downarrow}-z_{2C}^{\downarrow})]\}. \end{aligned} \tag{2.49}$$

This effective conductivity can be obtained by using F's, which are determined by the following boundary conditions.

Boundary Condition

On the surfaces of this structure electrons are reflected or diffusively scattered and we indicate the specular reflectivity of the surfaces by $p^{\uparrow,\downarrow}$, so that $(1\text{-}p^{\uparrow,\downarrow})$ is the coefficient of surface diffusive scattering. At the interfaces electrons transmit and are reflected or are diffusively scattered. We indicate the coefficient of transmission by $T^{\uparrow,\downarrow}$, the coefficient of reflection by $R^{\uparrow,\downarrow}$ and the coefficient of diffuse scattering by $D^{\uparrow,\downarrow}$ and $R^{\uparrow,\downarrow}+T^{\uparrow,\downarrow}+D^{\uparrow,\downarrow}=1$. On the hypothetical boundary (I) at the center of the SL quantization axes of the electron spins rotate by 2ϕ. These boundary conditions are

boundary	boundary condition	
(0)	$g_{\mathrm{A+}}^{\uparrow,\downarrow}=p_{\mathrm{A}}^{\uparrow,\downarrow}g_{\mathrm{A-}}^{\uparrow,\downarrow}$	(2.50)
(1)	$g_{\mathrm{A-}}^{\uparrow,\downarrow}=R_1^{\uparrow,\downarrow}g_{\mathrm{A+}}^{\uparrow,\downarrow}+T_1^{\uparrow,\downarrow}g_{\mathrm{B1-}}^{\uparrow,\downarrow}$	(2.51)
(1)	$g_{\mathrm{B1+}}^{\uparrow,\downarrow}=R_1^{\uparrow,\downarrow}g_{\mathrm{B1-}}^{\uparrow,\downarrow}+T_1^{\uparrow,\downarrow}g_{\mathrm{A+}}^{\uparrow,\downarrow}$	(2.52)
(I)	$g_{\mathrm{B1-}}^{\uparrow,\downarrow}=g_{\mathrm{B2-}}^{\uparrow,\downarrow}\cos^2\phi+g_{\mathrm{B2-}}^{\downarrow,\uparrow}\sin^2\phi$	(2.53)
(I)	$g_{\mathrm{B2+}}^{\uparrow,\downarrow}=g_{\mathrm{B1+}}^{\uparrow,\downarrow}\cos^2\phi+g_{\mathrm{B1+}}^{\downarrow,\uparrow}\sin^2\phi$	(2.54)
(2)	$g_{\mathrm{B2-}}^{\uparrow,\downarrow}=R_2^{\uparrow,\downarrow}g_{\mathrm{B2+}}^{\uparrow,\downarrow}+T_2^{\uparrow,\downarrow}g_{\mathrm{C-}}^{\uparrow,\downarrow}$	(2.55)
(2)	$g_{\mathrm{C+}}^{\uparrow,\downarrow}=R_2^{\uparrow,\downarrow}g_{\mathrm{C-}}^{\uparrow,\downarrow}+T_2^{\uparrow,\downarrow}g_{\mathrm{B2+}}^{\uparrow,\downarrow}$	(2.56)
(3)	$g_{\mathrm{C-}}^{\uparrow,\downarrow}=p_{\mathrm{C}}^{\uparrow,\downarrow}g_{\mathrm{C+}}^{\uparrow,\downarrow}$	(2.57)

Using (2.45), these conditions are expressed as follows:

boundary	boundary condition
(0)	$p_{\mathrm{A}}^{\uparrow,\downarrow}z_{0\mathrm{A}}^{\uparrow,\downarrow}\text{-}F_{\mathrm{A-}}^{\uparrow,\downarrow}(1/z_{0\mathrm{A}}^{\uparrow,\downarrow})F_{\mathrm{A+}}^{\uparrow,\downarrow}=1\text{-}p_{\mathrm{A}}^{\uparrow,\downarrow},$
(1)	$(R_1^{\uparrow,\downarrow}/z_{1\mathrm{A}}^{\uparrow,\downarrow})F_{\mathrm{A+}}^{\uparrow,\downarrow}\text{-}z_{1\mathrm{A}}^{\uparrow,\downarrow}F_{\mathrm{A-}}^{\uparrow,\downarrow}$ $+T_1^{\uparrow,\downarrow}y_{\mathrm{B/A}}^{\uparrow,\downarrow}z_{1\mathrm{B}}F_{\mathrm{B1-}}^{\uparrow,\downarrow}=1\text{-}R_1^{\uparrow,\downarrow}\text{-}T_1^{\uparrow,\downarrow}y_{\mathrm{B/A}}^{\uparrow,\downarrow},$
(1)	$(T_1^{\uparrow,\downarrow}y_{\mathrm{A/B}}^{\uparrow,\downarrow}/z_{1\mathrm{A}}^{\uparrow,\downarrow})F_{\mathrm{A+}}^{\uparrow,\downarrow}\text{-}(1/z_{1\mathrm{B}})F_{\mathrm{B1+}}^{\uparrow,\downarrow}$ $+R_1^{\uparrow,\downarrow}z_{1\mathrm{B}}F_{\mathrm{B1-}}^{\uparrow,\downarrow}=1\text{-}R_1^{\uparrow,\downarrow}\text{-}T_1^{\uparrow,\downarrow}y_{\mathrm{A/B}}^{\uparrow,\downarrow},$
(I)	$F_{\mathrm{B1-}}^{\uparrow,\downarrow}=F_{\mathrm{B2-}}^{\uparrow,\downarrow}\cos^2\phi+F_{\mathrm{B2-}}^{\downarrow,\uparrow}\sin^2\phi,$
(I)	$F_{\mathrm{B2+}}^{\uparrow,\downarrow}=F_{\mathrm{B1+}}^{\uparrow,\downarrow}\cos^2\phi+F_{\mathrm{B1+}}^{\downarrow,\uparrow}\sin^2\phi,$
(2)	$(R_2^{\uparrow,\downarrow}/z_{2\mathrm{B}})F_{\mathrm{B2+}}^{\uparrow,\downarrow}\text{-}z_{2\mathrm{B}}F_{\mathrm{B2-}}^{\uparrow,\downarrow}+T_2^{\uparrow,\downarrow}y_{\mathrm{C/B}}^{\uparrow,\downarrow}z_{3\mathrm{C}}^{\uparrow,\downarrow}F_{\mathrm{C-}}^{\uparrow,\downarrow}$ $=1\text{-}R_2^{\uparrow,\downarrow}\text{-}T_2^{\uparrow,\downarrow}y_{\mathrm{C/B}}^{\uparrow,\downarrow},$
(2)	$(T_2^{\uparrow,\downarrow}y_{\mathrm{B/C}}^{\uparrow,\downarrow}/z_{2\mathrm{B}})\mathrm{F}_{\mathrm{B2+}}^{\uparrow,\downarrow}\text{-}(1/z_{2\mathrm{C}}^{\uparrow,\downarrow})F_{\mathrm{C+}}^{\uparrow,\downarrow}$ $+R_2^{\uparrow,\downarrow}z_{2\mathrm{C}}^{\uparrow,\downarrow}F_{\mathrm{C-}}^{\uparrow,\downarrow}=1\text{-}R_2^{\uparrow,\downarrow}\text{-}T_2^{\uparrow,\downarrow}y_{\mathrm{B/C}}^{\uparrow,\downarrow},$
(3)	$(p_{\mathrm{C}}^{\uparrow,\downarrow}/z_{3\mathrm{C}}^{\uparrow,\downarrow})F_{\mathrm{C+}}^{\uparrow,\downarrow}\text{-}z_{3\mathrm{C}}^{\uparrow,\downarrow}F_{\mathrm{C-}}^{\uparrow,\downarrow}=1\text{-}p_{\mathrm{C}}^{\uparrow,\downarrow},$

where we write $y_{\mathrm{B/A}}^{\uparrow,\downarrow}=\lambda_{\mathrm{B}}/\lambda_{\mathrm{A}}^{\uparrow,\downarrow}$ and assume e, m and v_{F} are the same in all layers. These conditions are simultaneous linear equations in the F's, and can be solved in the usual way by giving appropriate values to λ, p, R and T (=1-D-R), respectively.

MR Ratio

Since the F's are functions of cos 2ϕ the conductivity σ_{eff} is given by the function of cos 2ϕ. This is the MR-effect in the basic structure and its MR ratio χ is given by

$$\chi = [\rho_{\mathrm{eff}}(\pi)-\rho_{\mathrm{eff}}(0)]/\ \rho_{\mathrm{eff}}(0) = [\sigma_{\mathrm{eff}}(0)-\sigma_{\mathrm{eff}}(\pi)]/\sigma_{\mathrm{eff}}(\pi), \tag{2.58}$$

where $\sigma_{\mathrm{eff}}(0)$ and $\sigma_{\mathrm{eff}}(\pi)$ correspond to the conductivities of parallel magnetization alignment (2ϕ=0) and of antiparallel alignment ($2\phi=\pi$).

Examples of Calculation

1 Symmetric Trilayer

Barnas et al.[7] calculated the MR ratio χ of a symmetric trilayer as functions of p and T with assuming R=0, i.e., diffuse scattering at interfaces.

Putting

$$h_{\mathrm{A}}^{\uparrow,\downarrow}=F_{\mathrm{A+}}^{\uparrow,\downarrow}(1/z_{0\mathrm{A}}^{\uparrow,\downarrow}-1/z_{1\mathrm{A}}^{\uparrow,\downarrow})+F_{\mathrm{A-}}^{\uparrow,\downarrow}(z_{1\mathrm{A}}^{\uparrow,\downarrow}-z_{0\mathrm{A}}^{\uparrow,\downarrow})$$
$$h_{\mathrm{C}}^{\uparrow,\downarrow}=F_{\mathrm{C+}}^{\uparrow,\downarrow}(1/z_{2\mathrm{C}}^{\uparrow,\downarrow}-1/z_{3\mathrm{C}}^{\uparrow,\downarrow})+F_{\mathrm{C-}}^{\uparrow,\downarrow}(z_{3\mathrm{C}}^{\uparrow,\downarrow}-z_{2\mathrm{C}}^{\uparrow,\downarrow})$$
$$h_{\mathrm{B}}=[(F_{\mathrm{B1+}}^{\uparrow}+F_{\mathrm{B1+}}^{\downarrow})(1/z_{1\mathrm{B1}}-1/z_{\mathrm{IB1}})+(F_{\mathrm{B1-}}^{\uparrow}+F_{\mathrm{B1-}}^{\downarrow})(z_{\mathrm{IB1}}-z_{1\mathrm{B1}})]$$
$$+[(F_{\mathrm{B2+}}^{\uparrow}+F_{\mathrm{B2+}}^{\downarrow})(1/z_{\mathrm{IB2}}-1/z_{2\mathrm{B2}})+(F_{\mathrm{B2-}}^{\uparrow}+F_{\mathrm{B2-}}^{\downarrow})(z_{2\mathrm{B2}}-z_{\mathrm{IB2}})]$$

and choosing the origin at (I), i.e., z_{I}=0, these h's become

$$h_{\mathrm{A}}^{\uparrow,\downarrow}=F_{\mathrm{A+}}^{\uparrow,\downarrow}(1/z_{0\mathrm{A}}^{\uparrow,\downarrow}-1/z_{1\mathrm{A}}^{\uparrow,\downarrow})+F_{\mathrm{A-}}^{\uparrow,\downarrow}(z_{1\mathrm{A}}^{\uparrow,\downarrow}-z_{0\mathrm{A}}^{\uparrow,\downarrow})$$
$$h_{\mathrm{C}}^{\uparrow,\downarrow}=F_{\mathrm{C+}}^{\uparrow,\downarrow}(z_{1\mathrm{A}}^{\uparrow,\downarrow}-z_{0\mathrm{A}}^{\uparrow,\downarrow})+F_{\mathrm{C-}}^{\uparrow,\downarrow}(1/z_{0\mathrm{A}}^{\uparrow,\downarrow}-1/z_{1\mathrm{A}}^{\uparrow,\downarrow})$$
$$h_{\mathrm{B}}=[(F_{\mathrm{B1+}}^{\uparrow}+F_{\mathrm{B1+}}^{\downarrow}+F_{\mathrm{B2-}}^{\uparrow}+F_{\mathrm{B2-}}^{\downarrow})(1/z_{1\mathrm{B1}}-1/z_{\mathrm{IB1}})+[(F_{\mathrm{B2+}}^{\uparrow}+F_{\mathrm{B2+}}^{\downarrow}+F_{\mathrm{B1-}}^{\uparrow}+F_{\mathrm{B1-}}^{\downarrow})(z_{\mathrm{IB1}}-z_{1\mathrm{B1}}).$$

In the case of a symmetric structure $t_{\mathrm{F}}=t_{\mathrm{A}}=t_{\mathrm{C}}$, $t=2t_{\mathrm{F}}+t_{\mathrm{B}}$. If the ferromagnetic layers A and B have the same electronic properties, i.e., $\alpha = \sigma_{\mathrm{A}}^{\uparrow}/\sigma_{\mathrm{A}}^{\downarrow}=\sigma_{\mathrm{C}}^{\uparrow}/\sigma_{\mathrm{C}}^{\downarrow}$, $\sigma_{\mathrm{F}}= \sigma_{\mathrm{A}}^{\uparrow}+\sigma_{\mathrm{A}}^{\downarrow} = \sigma_{\mathrm{C}}^{\uparrow}+\sigma_{\mathrm{C}}^{\downarrow}$, and $n_{\mathrm{F}}\lambda_{\mathrm{F}} = n^{\uparrow}\lambda^{\uparrow}+n^{\downarrow}\lambda^{\downarrow}$, then we put $\mu = \sigma_{\mathrm{B}}/\sigma_{\mathrm{F}}= (n_{\mathrm{B}}/n_{\mathrm{F}})(y_{\mathrm{B/F}})\{(e^2/mv_{\mathrm{F}})_{\mathrm{B}}/((e^2/mv_{\mathrm{F}})_{\mathrm{F}})\}$, $\eta = (\sigma_{\mathrm{A}}^{\uparrow}-\sigma_{\mathrm{A}}^{\downarrow})/(\sigma_{\mathrm{A}}^{\uparrow}+\sigma_{\mathrm{A}}^{\downarrow}) = (\alpha-1)/(\alpha+1)$, $\sigma^{\uparrow} = \sigma_{\mathrm{F}}(1+\eta)/2$ and $\sigma^{\downarrow}=\sigma_{\mathrm{F}}(1-\eta)/2$. The effective conductance is expressed as

$$\sigma_{\mathrm{eff}}\, t= [2\sigma_{\mathrm{F}}\, t_{\mathrm{F}} +\sigma_{\mathrm{B}}\, t_{\mathrm{B}}]+(3/4)\int_0^{\pi/2}\mathrm{d}\theta\,\sin^3\theta\cos\theta$$
$$\times[\sigma^{\uparrow}\lambda^{\uparrow}(h_{\mathrm{A}}^{\uparrow}+h_{\mathrm{C}}^{\uparrow})+\sigma^{\downarrow}\lambda^{\downarrow}(h_{\mathrm{A}}^{\downarrow}+h_{\mathrm{C}}^{\downarrow})+\sigma_{\mathrm{B}}\lambda_{\mathrm{B}}h_{\mathrm{B}}]. \tag{2.59}$$

This equation can be written as

$$\sigma_{\mathrm{eff}} = \sigma_{\mathrm{F}}\,\kappa(2\phi),$$
$$\kappa(2\phi)= 1+(\mu-1)t_{\mathrm{B}}/t$$
$$+(3/8\,t)\int_0^{\pi/2}\mathrm{d}\theta\,\sin^3\theta\cos\theta\,[(1+\eta)\lambda_{\mathrm{F}}^{\uparrow}h_{\mathrm{F}}^{\uparrow}+(1-\eta)\lambda_{\mathrm{F}}^{\downarrow}h_{\mathrm{F}}^{\downarrow}+2\mu\lambda_{\mathrm{B}}h_{\mathrm{B}}], \tag{2.60}$$

where $h_{\mathrm{F}}^{\uparrow,\downarrow}=h_{\mathrm{A}}^{\uparrow,\downarrow}+h_{\mathrm{C}}^{\uparrow,\downarrow}$. The MR effect is $\chi(2\phi)=\kappa(0)/\ \kappa(2\phi)-1$ and the MR ratio is $\chi(\pi)= \kappa(0)/\ \kappa(\pi)-1$.

In the case of diffuse interface scattering the Rs are zero and boundary conditions are given by

boundary	boundary condition
(0)	$p_A^{\uparrow,\downarrow} z_{0A}^{\uparrow,\downarrow} F_{A-}^{\uparrow,\downarrow} - (1/z_{0A}^{\uparrow,\downarrow}) F_{A+}^{\uparrow,\downarrow} = 1-p_A^{\uparrow,\downarrow}$
(1)	$-z_{1A}^{\uparrow,\downarrow} F_{A-}^{\uparrow,\downarrow} + T_1^{\uparrow,\downarrow} y_{B/A}^{\uparrow,\downarrow} z_{1B} F_{B1-}^{\uparrow,\downarrow} = 1-T_1^{\uparrow,\downarrow} y_{B/A}^{\uparrow,\downarrow}$
(1)	$(T_1^{\uparrow,\downarrow} y_{A/B}^{\uparrow,\downarrow} / z_{1A}^{\uparrow,\downarrow}) F_{A+}^{\uparrow,\downarrow} - (1/z_{1B}) F_{B1+}^{\uparrow,\downarrow} = 1-T_1^{\uparrow,\downarrow} y_{A/B}^{\uparrow,\downarrow}$
(I)	$F_{B1-}^{\uparrow,\downarrow} = F_{B2-}^{\uparrow,\downarrow} \cos^2\phi + F_{B2-}^{\downarrow,\uparrow} \sin^2\phi$
(I)	$F_{B2+}^{\uparrow,\downarrow} = F_{B1+}^{\uparrow,\downarrow} \cos^2\phi + F_{B1+}^{\downarrow,\uparrow} \sin^2\phi$
(2)	$-(1/z_{1B}) F_{B2-}^{\uparrow,\downarrow} + T_2^{\uparrow,\downarrow} y_{C/B}^{\uparrow,\downarrow} (1/z_{0A}^{\uparrow,\downarrow}) F_{C-}^{\uparrow,\downarrow} = 1-T_2^{\uparrow,\downarrow} y_{C/B}^{\uparrow,\downarrow}$
(2)	$(T_2^{\uparrow,\downarrow} y_{B/C}^{\uparrow,\downarrow} z_{1B}) F_{B2+}^{\uparrow,\downarrow} - z_{1A} F_{C+}^{\uparrow,\downarrow} = 1-T_2^{\uparrow,\downarrow} y_{B/C}^{\uparrow,\downarrow}$
(3)	$(p_C^{\uparrow,\downarrow} z_{0A}^{\uparrow,\downarrow}) F_{C+}^{\uparrow,\downarrow} - (1/z_{0A}^{\uparrow,\downarrow}) F_{C-}^{\uparrow,\downarrow} = 1-p_C^{\uparrow,\downarrow}$

The treatment of the MR ratio mentioned in this section can extend directly to other metal multilayers, such as, dual spin-valve structures etc.

2 Spin Valve with Capping Layer

Sakakima *et al.* [8] have studied the MR effect of a spin valve with a capping layer for improving the MR ratio by in creasing the specular reflection p of its surface. They measured the MR ratios in a spin valve having the layer structure of [buffer/FeNi(5nm)/Cu(2.8nm)/Co(5nm)/Ag-capping], by adjusting the thickness t_{Ag} of the Ag capping layer. They calculated the MR ratio of a spin valve composed of five layers including a buffer layer and a capping layer shown in Fig. 2.16(a).

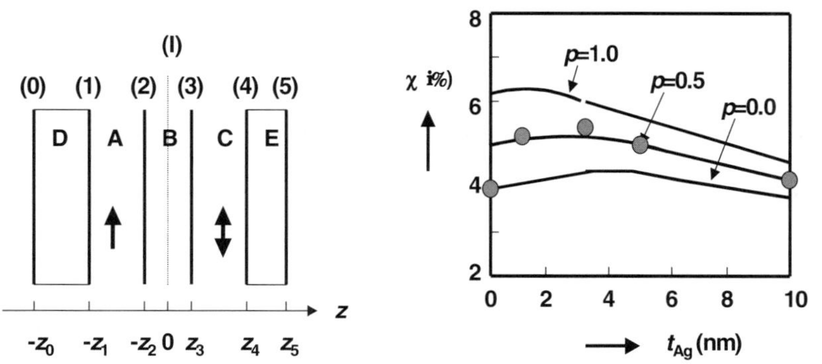

(a) Calculation model

(b) Calculated results (solid line) and experimental data (circle)

Fig. 2.16. (**a**) Layer structure for the calculation, D is a buffer layer and E is a capping layer and (**b**) comparison between experimental data and calculated results of MR ratios in a spin valve with Ag capping layer.

In this model the effective conductance is

$$\sigma_{\text{eff}} = \sigma_F \kappa(2\phi),$$

$$\kappa(2\phi) = 1+(\mu-1)(t_B+t_D+\mathbf{t}_E)/t+(3/4t\sigma_F)\int^{\pi/2}_0 d\theta \sin^3\theta \cos\theta$$

$$\times\{\sigma_A[(1+\eta)\lambda_A^{\uparrow} h_A^{\uparrow} + (1-\eta)\lambda_A^{\downarrow} h_A^{\downarrow}]+\sigma_C[(1+\eta)\lambda_C^{\uparrow} h_C^{\uparrow} + (1-\eta)\lambda_C^{\downarrow} h_C^{\downarrow}]$$

$$+\sigma_B\lambda_B h_B+\sigma_D\lambda_D h_D +\sigma_E\lambda_E h_E\}, \tag{2.61}$$

where $\sigma_F=(\sigma_A t_A+\sigma_C t_C)/(t_A+t_C)$, $\sigma_A=\sigma_A^{\uparrow}+\sigma_A^{\downarrow}$ etc., $\mu=(\sigma_B t_B+\sigma_D t_D+\sigma_E t_E)/(t_B+t_D+t_E)\sigma_F$, here the h's are similar parameters determined by the boundary conditions mentioned in the above example. Assuming $T^{\uparrow}=0.8$ and $T^{\downarrow}=0.2$, the MR ratio is estimated as functions of p and t_{Ag}. Their results are summarized in Fig. 2.16(b), which shows that the MR ratio χ is improved by adding an adequate reflection layer on spin valves with improved specularity p of 0.5. Details are given in Chapter 4.

2.3 Free-Electron Model of TMR Devices

A tunnel junction, which comprises an insulator film sandwiched between two electrode films, is discussed with the tunnel current caused by an electric voltage applied to the insulator film. The first description is of Simmons' formula [9], which expresses the tunnel current as a function of the potential barrier in an insulator film (Sect. 2.3.1). The second description is of a ferromagnetic tunnel junction, in which the two electrodes are ferromagnetic metal films. The tunnel current and the TMR effect are explained according to Slonczewski [10] (Sect. 2.3.2).

2.3.1 Simmons' Formula for the Tunnel Current

A tunnel junction consists of metal electrodes (1) and (2), which are separated from each other by an insulator film (I) as shown in Fig.2.17. An electric voltage V is applied across the junction (along the x-axis) from electrode (2) to electrode (1). The energies of the electrons in electrode (2) are lowered by eV relative to the energies in electrode (1). The electric current J flows from electrodes (2) to (1) due to electron tunneling through the barrier in the insulator film, which has thickness t and barrier potential $V(x)$. $V(x)$ crosses the Fermi level ε_F of electrode (1) at x_1 and x_2. The number n_1 of electrons tunneling through the insulator film from electrode (1) to electrode (2) is

$$n_1 = \int^{vF}_0 n(v_x)T(\varepsilon_x)\, dv_x=(1/m)\int^{\varepsilon F}_0 D(\varepsilon_x)T(\varepsilon_x)d\varepsilon_x, \tag{2.62}$$

where $\varepsilon_x=(mv_x^2)/2$, v_x is the x-component of the electron velocity $\boldsymbol{v}$ and $n(v_x)$ is the number of electrons per unit volume with velocity between v_x and v_x+dv_x. The state density $D(\varepsilon_x)$ for electrons with ε_x is given by

$$D(\varepsilon_x) = (2m^3/h^3)\iint^{\infty}_{-\infty} dv_y dv_z f_0(\varepsilon) = (4\pi m^2/h^3)\int^{\infty}_0 f_0(\varepsilon_r+\varepsilon_x)d\varepsilon_r. \tag{2.63}$$

Here we use cylindrical coordinates and $v_r^2 = v_y^2 + v_z^2$ and $\varepsilon_r = \mathbf{v}_r^2/2m$. Therefore,

$$n_1 = (4\pi m^2/h^3)\int^{\varepsilon F}_0 T(\varepsilon_x)\, d\varepsilon_x \int^{\infty}_0 f_0(\varepsilon_r+\varepsilon_x) d\varepsilon_r. \qquad (2.64)$$

While the tunneling probability $T(\varepsilon_x)$ is calculated by the WKB approximation as follows:

$$T(\varepsilon_x) = \exp[-(4\pi/h)\int^{x2}_{x1}\{2m(V(x)- \varepsilon_x)\}^{1/2} dx]. \qquad (2.65)$$

Likewise, the number n_2 of electrons tunneling through the insulator film from electrode (2) to electrode (1) is

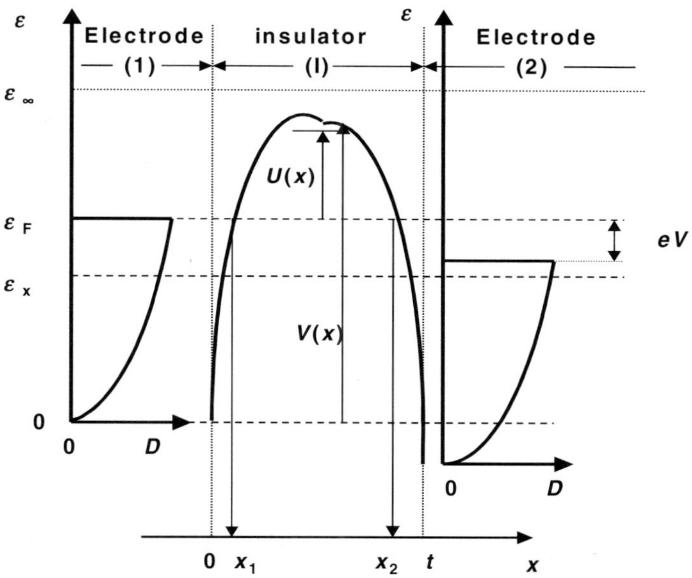

Fig. 2.17. Potential barrier in tunnel junction (I) and state densities in metal electrodes (1) and (2)

$$n_2 = (4\pi m^2/h^3)\int^{\varepsilon F}_0 T(\varepsilon_x)\, d\varepsilon_x \int^{\infty}_0 f_0(\varepsilon_r+\varepsilon_x+eV) d\varepsilon_r. \qquad (2.66)$$

The net number n of electrons tunneling through the insulator film is

$$n = n_1 - n_2 = (4\pi m^2/h^3)\int^{\varepsilon F}_0 T(\varepsilon_x)\, d\varepsilon_x \int^{\infty}_0 [f_0(\varepsilon_r+\varepsilon_x) - f_0(\varepsilon_r+\varepsilon_x+eV)] d\varepsilon_r. \qquad (2.67)$$

The tunnel current J is

$$J = -en = -(4\pi e m^2/h^3)\int^{\varepsilon F}_0 T(\varepsilon_x) d\varepsilon_x \int^{\infty}_0 [f_0(\varepsilon_r+\varepsilon_x) - f_0(\varepsilon_r+\varepsilon_x+eV)] d\varepsilon_r. \qquad (2.68)$$

If we express the potential barrier as $V(x)=\varepsilon_F+U(x)$, then $T(\varepsilon_x)$ is given approximately by

$$T(\varepsilon_x) \cong \exp[-A(\varepsilon_F+U_m-\varepsilon_x)]^{1/2}. \qquad (2.69)$$

Here $A=(4\pi\beta w/h)(2m)^{1/2}$ with $U_m=\int^{x2}_{x1}U(x)^{1/2}dx/w$ (mean barrier height) and $\beta=1-(1/8U_m^2w)\int^{x2}_{x1}[U(x)-U_m]^2dx$ and $w=x_2-x_1$. At 0K the integral including the Fermi functions is

$$\int^{\infty}_0[f_0(\varepsilon_r+\varepsilon_x)-f_0(\varepsilon_r+\varepsilon_x+eV)]d\varepsilon_r = -eV, \qquad 0<\varepsilon_x<\varepsilon_F-eV,$$
$$=\varepsilon_F-eV, \qquad \varepsilon_F-eV<\varepsilon_x<\varepsilon_F$$
$$= 0. \qquad \varepsilon_F<\varepsilon_x \tag{2.70}$$

Thus, the tunnel current J becomes

$$J = (4\pi em^2/h^3)\{\int^{\varepsilon F-eV}_0 eV\exp[-A(\varepsilon_F+U_m-\varepsilon_x)]^{1/2}d\varepsilon_x$$
$$+\int^{\varepsilon F}_{\varepsilon F-eV}(\varepsilon_F-eV)\exp[-A(\varepsilon_F+U_m-\varepsilon_x)]^{1/2}d\varepsilon_x\}. \tag{2.71}$$

This current can be expressed by the U_m and the width w as follows:

$$J = J_0\{U_m\exp(-AU_m^{1/2})-(U_m+eV)\exp[-A(U_m+eV)^{1/2}]\}, \tag{2.72}$$

with $J_0=e/2\pi h(\beta w)^2$ (Simmons' formula). The first term is due to tunneling electrons from (1) to (2) and the second term is due to electrons from (2) to (1).

In the case of $eV<<U_m$

$$J.= [(2m)^{1/2}/w](e/h)^2U_m^{1/2}\exp(-AU_m^{1/2})V. \tag{2.73}$$

This shows Ohm's law and the conductance σ_T is given by

$$\sigma_T= [(2m)^{1/2}/w](e/h)^2U_m^{1/2}\exp(-AU_m^{1/2}). \tag{2.74}$$

This low-voltage performance of the tunnel junction is important for the analysis of the TMR effect.

2.3.1.1 Tunnel Current for a Square Potential

The square potential has constant height U_0 and $w=t$ as shown in Fig. 2.18. In the case of low applied voltage $V\sim0, U_m=U_0$ and the tunnel current is

$$J = [(2m)^{1/2}/t](e/h)^2U_0^{1/2}\exp(-AU_0^{1/2})V = \sigma_TV. \tag{2.75}$$

In the intermediate potential $U_m>eV$, $U_m=(U_0-eV/2)$ and J is

$$J = J_0\{(U_0-eV/2)\exp[-A(U_0-eV/2)^{1/2}]$$
$$-(U_0+eV/2)\exp[-A(U_0+eV/2)^{1/2}]\} \tag{2.76}$$

where $J_0=-e/2\pi h(\beta t)^2$, $A=(4\pi\beta t/h)(2m)^{1/2}$ with correction factor $\beta=1-\{(eV)^2/96\}(U_0+\varepsilon_F-\varepsilon_x-eV/2)$.

Image Force Potential

An electron in the insulator has its own image in both electrodes and receives image forces. The image force potential V_i is given by

$$V_i(x) = (-e^2/4\pi\kappa)[1/2x+\sum_{n=1}^{\infty}\{nt/[(nt)^2-x^2]-1/nt\}], \tag{2.77}$$

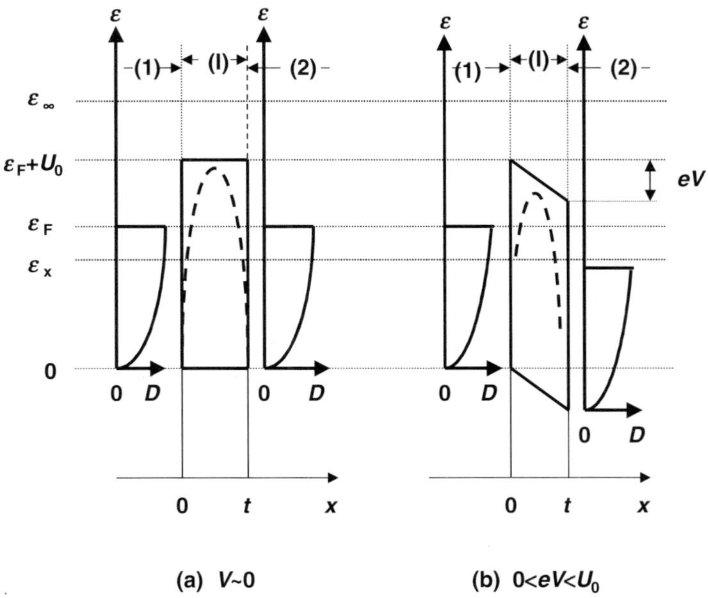

Fig. 2.18. Square potential barrier: (**a**) low-voltage range, (**b**) intermediate range, U0 is the barrier height and t is the width of the barrier. The potential of the dotted line shows the image potential

where κ is the dielectric constant of the insulator film. Instead of handling this troublesome equation Simmons took the approximation:

$$V_i(x) = -1.15\gamma t^2/x(t-x), \tag{2.78}$$

Here $\gamma = V_i(t/2)$ and is given by

$$\gamma = (-e^2/2\pi\kappa)\Sigma_{n=1}^{\infty}(-1)^n/n = (-e^2/2\pi\kappa)\ \ln 2. \tag{2.79}$$

Using this approximation, the barrier potential becomes

$$U(x) = U_0 - eVx/t - 1.15\gamma t^2/x(t-x), \tag{2.80}$$

and is shown in Fig. 2.18. with dotted lines.
The mean barrier potential U_m is

$$\begin{aligned} U_m &= \{\textstyle\int_{t1}^{t2}[U_0 - eVx/t - 1.15\gamma t^2/x(t-x)]\ dx\}/w \\ &= U_0 - (eV/2)(t_1+t_2) - [1.15\gamma t/(t_2-t_1)]\ln[t_2(t-t_1)/t_1(t-t_2)], \end{aligned} \tag{2.81}$$

where t_1 and t_2 are roots of $U_0 - eVx/t - 1.15\gamma t^2/x(t-x) = 0$.
In the intermediate potential ($U_0 > eV$) these roots are given approximately by;

$$t_1 = 1.2\gamma t/U_0,\ t_2 = t[1 - 9.2\gamma/(3U_0 + 4\gamma - 2eV)] + t_1, \tag{2.82}$$

Using the barrier height U_m and width $w(=t_2 - t_1)$, the tunnel current J is calculated from Simmons' formula (2.72).

2.3.1.2 Tunnel Current Expressions in Practical Units

Simmons gave the tunnel current expressions in practical units of J in A/cm^2, potential in V, and the sizes in A. For the convenience of readers, his expressions are listed:

[General potential barrier]

$$J = [6.2\times10^{10}/(\beta w)^{1/2}]\{U_m\exp(-1.025\beta wU_m^{1/2}) -(U_m+V)\exp[-1.025\beta w(U_m+V)^{1/2}]\}, \tag{2.83}$$

and in the case of low applied voltage (V~0)

$$J = 3.16\times10^{10}U_m^{1/2}\exp(-1.025wU_m^{1/2})V/w. \tag{2.84}$$

[Square potential barrier]

$U_0>V>0$:

$$J = (6.2\times10^{10}/w^{1/2})\ \{(U_0-V/2)\exp[-1.025w(U_0-V/2)^{1/2}] -(U_0+V/2)\exp[-1.025w(U_0+V/2)^{1/2}]\}, \tag{2.85}$$

and the low-voltage case (V~0)

$$J = 3.16\times10^{10}U_0^{1/2}\exp(-1.025wU_0^{1/2})V/w. \tag{2.86}$$

2.3.2 Ferromagnetic Tunnel Junction

Simple Model for TMR Ratio

The ferromagnetic tunnel junction is composed of two ferromagnetic electrodes (1) and (2) separated by an insulator film (I). We consider the tunnel current in this junction assuming the separator film (I) has a square barrier potential $U(x)$ and its thickness is t, and the magnetizations of both electrodes are $\boldsymbol{M}_1$ and $\boldsymbol{M}_2$. If a voltage is applied between electrode (2) and (1), tunnel currents flow from (2) to (1) as discussed in the previous section Sect. 2.3.1, but at low temperature in ferromagnetic tunnel junctions these currents consist of two components, i.e., due to ↑-spin electrons and ↓-spin electrons (two-current model).

First we explain a simple model of the MR effect in ferromagnetic junctions in a low applied voltage V~0. Fig. 2.19. shows electron energy configurations in the junction. The Fermi level ε_F of electrode (1) is lowered by eV relative to that of electrode (2) and the barrier height U_0 is higher than the Fermi level ε_F. The conduction electrons have approximately kinetic energies of the ε_F and tunnel through the barrier, conserving their spins. Therefore, in the case of parallel magnetizations of both electrodes (Fig. 2.19(a)) the electric conductance of the ferromagnetic junction is given approximately by

$$\sigma_T^{\uparrow\uparrow} \propto \exp(-AU_0^{1/2})\ [D_1^{\uparrow}(\varepsilon_F)D_2^{\uparrow}(\varepsilon_F+eV)+D_1^{\downarrow}(\varepsilon_F)D_2^{\downarrow}(\varepsilon_F+eV)], \tag{2.87}$$

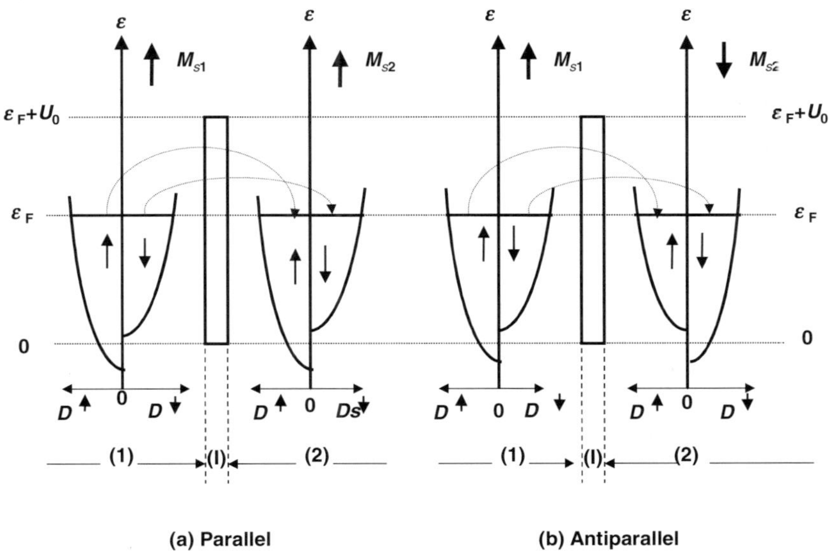

Fig. 2.19. Schematic presentation of the electron tunneling in a ferromagnetic junction: (**a**) magnetizations of both electrodes are in parallel case, and (**b**) are in antiparalel case

where the exponential term shows the tunneling probability of the electrons and $A=(4\pi w/\mathrm{h})(2m)^{1/2}$, and $D_{12}^{\uparrow}$ and $D_{12}^{\downarrow}$ are state densities for majority and minority spin electrons. For the antiparallel case of reversed magnetization $\boldsymbol{M}_2$ (Fig. 2.19(b)) ↑-spin electrons go through the barrier from the majority band of electrode (1) to the minority band of electrode (2) and ↓-spin electrons of the minority band of (1) go through the majority band of (2) and therefore the tunneling conductance is

$$\sigma_{\mathrm{T}}^{\uparrow\downarrow} \propto \exp(-AU_0^{1/2})[D_1^{\uparrow}(\varepsilon_{\mathrm{F}})D_2^{\downarrow}(\varepsilon_{\mathrm{F}}+eV)+D_1^{\downarrow}(\varepsilon_{\mathrm{F}})D_2^{\uparrow}(\varepsilon_{\mathrm{F}}+eV)]. \quad (2.88)$$

Thus the TMR ratio χ is given by

$$\chi \equiv (\sigma_{\mathrm{T}}^{\uparrow\uparrow}-\sigma_{\mathrm{T}}^{\uparrow\downarrow})/\,\sigma_{\mathrm{T}}^{\uparrow\downarrow}= 2(D_1^{\uparrow}-D_1^{\downarrow})(D_2^{\uparrow}-D_2^{\downarrow})/(D_1^{\uparrow}D_2^{\downarrow}+D_1^{\downarrow}D_2^{\uparrow})$$

$$= 2P_1P_2/(1-P_1P_2), \quad (2.89)$$

with $P_i=(D_i^{\uparrow}-D_i^{\downarrow})/(D_i^{\uparrow}+D_i^{\downarrow})$, i=1,2 and is determined by the polarizations P_1 and P_2 of the ferromagnetic electrodes. This relations is given by Julliere in his early work [11], and later Maekawa *et al.*[12] gave following equivalent expression:

$$\chi = 2(\sigma_{\mathrm{T}}^{\uparrow\uparrow}-\sigma_{\mathrm{T}}^{\uparrow\downarrow})/(\sigma_{\mathrm{T}}^{\uparrow\uparrow}+\sigma_{\mathrm{T}}^{\uparrow\downarrow\downarrow})=2P_1P_2.$$

Figure 2.20 compares this model with experimental TMR ratios for various ferromagnetic junctions. The experimental values are considerably lower than the value given by this model and the origin of these discrepancies is thought to be in the film quality of the thin insulator but another reason is discussed in the next section.

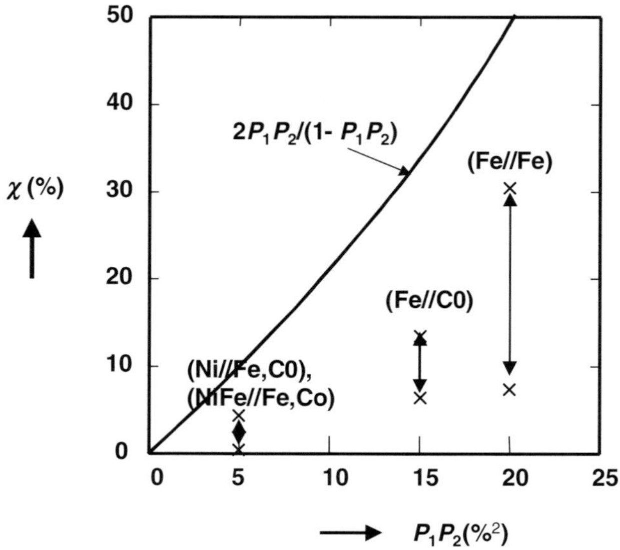

Fig. 2.20. Comparison between simple model and experimental values of the TMR ratio: the solid line shows the simple model and ranges indicated by arrows show scattering ranges of experimental values measured at 4 K for (Fe//Fe), (Fe//Co) and (Ni or Ni base alloys//Co) junctions

Physics of Ferromagnetic Tunnel Junctions

This section describes the TMR effect in a ferromagnetic tunnel junction in accordance with Slonchewski [10]. The junction has a square barrier potential of height U_0 and of width w and magnetizations $\boldsymbol{M}_1$ and $\boldsymbol{M}_2$ of both electrodes (1) and (2) tilt relative to each other by 2ϕ. We consider an applied electric voltage V sufficiently low, i.e., $eV<<\varepsilon_F$. At low temperature, tunnel currents are assumed to be carried by ↑-spin and ↓-spin electrons, independently (two-current model). The electron energy sheme for a ferromagnetic tunnel junction is shown in Fig. 2.21.

The numbers of electrons with ↑-spin and ↓-spin in ferromagnetic electrode (1) moving along the x-axis from the left into the potential barrier of the insulator film (I) and tunneling through the barrier to electrode (2) are $n_1^{\uparrow}$ and $n_1^{\downarrow}$, and the numbers of electrons from the right electrode (2) to the left electrode (1) are $n_2^{\uparrow}$ and $n_2^{\downarrow}$, respectively. $n_j^{\uparrow}$ and $n_j^{\downarrow}$ (j=1,2) are expressed as follows:

$$n_j^{\uparrow} = (4\pi m^2/\mathrm{h}^3)\int_{-JM_j}^{\varepsilon_{Fj}} T_j^{\uparrow}(\varepsilon_x)\mathrm{d}\varepsilon_x\int_{-JM_j}^{\infty} f_0(\varepsilon_r+\varepsilon_x)\mathrm{d}\varepsilon_r, \tag{2.90}$$

$$n_j^{\downarrow} = (4\pi m^2/\mathrm{h}^3)\int_{+JM_j}^{\varepsilon_{Fj}} T_j^{\downarrow}(\varepsilon_x)\mathrm{d}\varepsilon_x\int_{+JM_j}^{\infty} f_0(\varepsilon_r+\varepsilon_x)\mathrm{d}\varepsilon_r, \tag{2.91}$$

here j=1,2.

The tunnel current J is calculated as $J=J^{\uparrow}+J^{\downarrow}$ with $J^{\uparrow}=-e(n_1^{\uparrow}-n_2^{\uparrow})$, $J^{\downarrow}=-e(n_1^{\downarrow}-n_2^{\downarrow})$.The tunnel probabilities $T_1^{\uparrow,\downarrow}$ and $T_2^{\uparrow,\downarrow}$ are obtained by the following procedure which we show only for $T_1^{\uparrow}$ and others are obtained by symmetric transformation between suffixs 1 and 2 or ↑ and ↓.

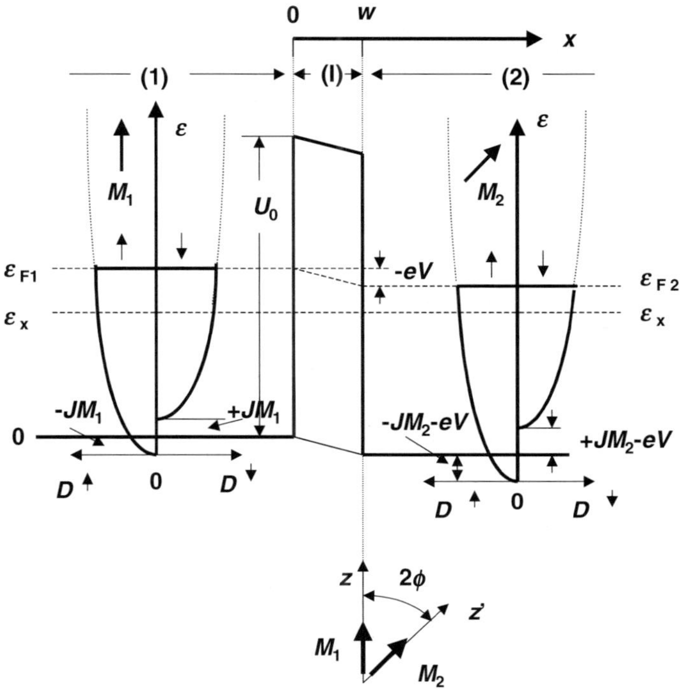

Fig.2.21. Energy sheme for a ferromagnetic junction

Electrons with ↑-spin come from the left to the tunnel barrier and tunnel through the barrier to the right electrode (2), This electron behavior is treated by the following effective Hamiltonian:

$$H = -(h^2/2m)(\mathrm{d}/\mathrm{d}x)^2+U(x)-eV-(\boldsymbol{H}_{ex}(x)\cdot\boldsymbol{\sigma}), \tag{2.92}$$

Here h (=h/2π) is the Dirac h, $\boldsymbol{\sigma}$ is the Pauli spin operator and $\boldsymbol{H}_{ex}(x)=J\boldsymbol{M}$ is effective exchange field in the molecular-field approximation, namely, the Hamiltonians are in electrode (1), $x\leq 0$, and electrode (2), $w\leq x$,

$$x\leq 0 \quad H= -(h^2/2m)(\mathrm{d}/\mathrm{d}x)^2-J_1(\boldsymbol{M}_1(x)\cdot\boldsymbol{\sigma}) \tag{2.93}$$

$$w\leq x \quad H= -(h^2/2m)(\mathrm{d}/\mathrm{d}x)^2-eV-J_2(\boldsymbol{M}_2(x)\cdot\boldsymbol{\sigma}), \tag{2.94}$$

and in the insulator (I),

$$0\leq x\leq w \quad H= -(h^2/2m)(\mathrm{d}/\mathrm{d}x)^2+U_0-eVx/w. \tag{2.95}$$

The wave functions in electrodes (1) and (2) are given by

$$x\leq 0 \quad \psi_1^{\uparrow,\downarrow}(x)= (m/hk_1^{\uparrow})^{-1/2}\exp(\mathrm{i}k_1^{\uparrow}x)+R^{\uparrow,\downarrow}\exp(-\mathrm{i}k_1^{\uparrow,\downarrow}x) \tag{2.96}$$

$$w\leq x \quad \psi_2^{\uparrow,\downarrow}(x)= \mathrm{C}^{\uparrow,\downarrow}\exp[\mathrm{i}k_2^{\uparrow,\downarrow}(x-w)], \tag{2.97}$$

with electron momenta

$$hk_1^{\uparrow,\downarrow}= [2m(\varepsilon_x+\sigma J_1 M_1)]^{1/2}, \tag{2.98}$$

$$hk_2^{\uparrow,\downarrow}= [2\mathrm{m}(\varepsilon_x+eV+\sigma J_2 M_2)]^{1/2}, \tag{2.99}$$

where $\sigma=\pm$(or $\uparrow,\downarrow$). If $\varepsilon_F > H_{exj}$ ($j = 1,2$), the Fermi wave vector $k_{Fj}^{\uparrow,\downarrow}$ is real. In the insulator the wave functions are

$$0 \le x \le w \quad \psi_I^{\uparrow,\downarrow}(x) = A^{\uparrow,\downarrow}\exp(-kx) + B^{\uparrow,\downarrow}\exp(kx), \tag{2.100}$$

with the imaginary momentum

$$\mathrm{i}\, hk = [-2m(U_0 - \varepsilon_x)]^{1/2}. \tag{2.101}$$

The eight constants in the wave functions $R^{\uparrow,\downarrow}$, $A^{\uparrow,\downarrow}$, $B^{\uparrow,\downarrow}$ and $C^{\uparrow,\downarrow}$ are determined by the following boundary conditions.

The wave functions in electrode (1) and in insulator (I) connect smoothly at their boundary, namely, the ψ's and $d\psi/dx$'s are equal at x=0. These conditions give the following relations:

$$k_1^{\uparrow -1/2} + R^{\uparrow} = A^{\uparrow} + B^{\uparrow} \tag{2.102}$$

$$\mathrm{i}\, k_1^{\uparrow}(k_1^{\uparrow -1/2} - R^{\uparrow}) = -k(A^{\uparrow} - B^{\uparrow}) \tag{2.103}$$

$$R^{\downarrow} = A^{\downarrow} + B^{\downarrow} \tag{2.104}$$

$$-\mathrm{i}\, k_1^{\downarrow} R^{\downarrow} = -k(A^{\downarrow} - B^{\downarrow}). \tag{2.105}$$

At $x=w$ the quantization axes change their direction by 2ϕ, and this condition requires spinor transformation to the ψ and $d\psi$/dx. The boundary conditions are

$$A^{\uparrow}\exp(-kw) + B^{\uparrow}\exp(kw) = C^{\uparrow}\cos\phi + C^{\downarrow}\sin\phi, \tag{2.106}$$

$$A^{\downarrow}\exp(-kw) + B^{\downarrow}\exp(kw) = -C^{\uparrow}\sin\phi + C^{\downarrow}\cos\phi, \tag{2.107}$$

$$-k\{A^{\uparrow}\exp(-kw) - B^{\uparrow}\exp(kw)\} = \mathrm{i}k_2^{\uparrow}C^{\uparrow}\cos\phi + \mathrm{i}k_2^{\downarrow}C^{\downarrow}\sin\phi, \tag{2.108}$$

$$-k\{A^{\downarrow}\exp(-kw) - B^{\downarrow}\exp(kw)\} = -\mathrm{i}k_2^{\uparrow}C^{\uparrow}\sin\phi + \mathrm{i}k_2^{\downarrow}C^{\downarrow}\cos\phi. \tag{2.109}$$

Solving these equations, we can obtain the following constants:

$$A^{\uparrow} = [-2\mathrm{i}k_1^{\uparrow 1/2} + (k + \mathrm{i}k_1^{\uparrow})B^{\uparrow}]/(k - \mathrm{i}k_1^{\uparrow}), \tag{2.110}$$

$$B^{\uparrow} = \{-2\mathrm{i}k_1^{\uparrow 1/2}[k^2 + k_2^{\uparrow}k_2^{\downarrow} + \mathrm{i}k(k_2^{\uparrow} - k_2^{\downarrow})\cos 2\phi]e^{-2kw}\}/ (k - \mathrm{i}k_1^{\uparrow})(k - \mathrm{i}k_2^{\uparrow})(k - \mathrm{i}k_2^{\downarrow}), \tag{2.111}$$

$$A^{\downarrow} = [(k + \mathrm{i}k_1^{\downarrow})B^{\downarrow}]/(k - \mathrm{i}k_1^{\downarrow}), \tag{2.112}$$

$$B^{\downarrow} = [2kk_1^{\uparrow 1/2}(k_2^{\downarrow} - k_2^{\uparrow})e^{-2kw}\sin 2\phi]/ (k - \mathrm{i}k_1^{\uparrow})(k - \mathrm{i}k_2^{\uparrow})(k - \mathrm{i}k_2^{\downarrow}), \tag{2.113}$$

$$C^{\uparrow} = [-4\mathrm{i}kk_1^{\uparrow 1/2}e^{-kw}\cos\phi]/(k - \mathrm{i}k_1^{\uparrow})(k - \mathrm{i}k_2^{\uparrow}), \tag{2.114}$$

$$C^{\downarrow} = [-4\mathrm{i}kk_1^{\uparrow 1/2}e^{-kw}\sin\phi]/(k - \mathrm{i}k_1^{\uparrow})(k - \mathrm{i}k_2^{\downarrow}). \tag{2.115}$$

The tunneling probability for the electron flux coming from the left side with spin $\uparrow$ is given by the following probability current density:

$$T_1^{\uparrow} = \mathrm{Re}(h/\mathrm{i}m)\Sigma_\sigma \psi^{\sigma *}[d\psi^{\sigma}/dx]. \tag{2.116}$$

Evaluations of $T_1^{\uparrow}$ are carried out at $x=0_+$, i.e., inside the barrier (I) ($0 \le x \le w$), in which the effective exchange field is assumed to be zero. When w is large (but smaller than the coherence length λ_φ of electrons), $|\psi_I^{\downarrow}/\psi_I^{\uparrow}| \cong e^{-2kw}$ becomes small and thus the contribution of $\downarrow$-spin electrons to $T_1^{\uparrow}$ is small. In this case $T_1^{\uparrow}$ is given by

$$\begin{aligned} T_1^{\uparrow}(x=0_+) &= \mathrm{Re}\{(h/\mathrm{i}m)\, \psi_I^{\uparrow *}[d\psi_I^{\uparrow}/dx]\} = \mathrm{Re}(-kh/\mathrm{i}m)(A^{\uparrow} + B^{\uparrow})^*(A^{\uparrow} - B^{\uparrow}) \\ &= \{8hk^2k_1^{\uparrow}e^{-2kw}(k^2 + k_2^{\uparrow}k_2^{\downarrow})(k_2^{\uparrow} + k_2^{\downarrow})/m(k^2 + k_1^{\uparrow 2})(k^2 + k_2^{\uparrow 2})(k^2 + k_2^{\downarrow 2})\} \\ &\quad \times [1 + P_{1\mathrm{eff}}^{\uparrow}\cos 2\phi]. \end{aligned} \tag{2.117}$$

with $P_{1eff}^{\uparrow}=[(k_2^{\uparrow}-k_2^{\downarrow})(k^2-k_2^{\uparrow}k_2^{\downarrow})/(k_2^{\uparrow}+k_2^{\downarrow})(k^2+k_2^{\uparrow}k_2^{\downarrow})]$. For electrons with $\downarrow$-spin the tunneling probability $T_1^{\downarrow}$ is obtained by replacing $k_1^{\uparrow}$ with $k_1^{\downarrow}$ and the total tunnel probability $T=T_1^{\uparrow}+T_1^{\downarrow}$ is given by

$$T=(8hk^2e^{-2kw}/m)\,\Pi_{j=1,2}[(k_j^{\uparrow}+k_j^{\downarrow})(k^2+k_j^{\uparrow}k_j^{\downarrow})/(k^2+k_j^{\uparrow 2})(k^2+k_j^{\downarrow 2})]$$
$$\times\{1+P_{1eff}^{\uparrow}P_{2eff}^{\uparrow}\cos 2\phi\}, \tag{2.118}$$

with effective polarization

$$P_{jeff}^{\uparrow}=[(k_j^{\uparrow}-k_j^{\downarrow})(k^2-k_j^{\uparrow}k_j^{\downarrow})/(k_j^{\uparrow}+k_j^{\downarrow})(k^2+k_j^{\uparrow}k_j^{\downarrow})].$$

Tunnel currents are calculated in a similar way to Sect. 2.3.1: electrons near the Fermi surface contribute these currents therefore the wave vectors $k(\varepsilon_x)$ are replaced by $k(\varepsilon_F)$. In the case of small voltage $V\sim 0$ the tunnel current J is expressed as follows (see 2.75):

$$J=\{[(2m)^{1/2}/w]\,(e/\mathrm{h})^2\}TV=\sigma_T V. \tag{2.119}$$

Therefore, the conductance σ_T of a ferromagnetic tunnel junction is

$$\sigma_T=\{[(2m)^{1/2}/w](e/\mathrm{h})^2\}T$$
$$=\sigma_0\{1+P_{1eff}^{\uparrow}P_{2eff}^{\uparrow}\cos 2\phi\}, \tag{2.120}$$

The MR ratio χ is given by

$$\chi=\sigma_F(2\phi=0)/\,\sigma_{AF}(2\phi=\pi)-1=2P_{1eff}P_{2eff}/(1-P_{1eff}P_{2eff}). \tag{2.121}$$

This shows Julliere's relation (2.89) to be modified by the effective polarizations, which relate to the polarizations as follows:

$$\begin{aligned}P_{jeff}^{\uparrow}&=[(k_j^{\uparrow}-k_j^{\downarrow})(k^2-k_j^{\uparrow}k_j^{\downarrow})/(k_j^{\uparrow}+k_j^{\downarrow})(k^2+k_j^{\uparrow}k_j^{\downarrow})]\\&=[(D_j^{\uparrow}-D_j^{\downarrow})/(D_j^{\uparrow}+D_j^{\downarrow})][(k^2-k_j^{\uparrow}k_j^{\downarrow})/(k^2+k_j^{\uparrow}k_j^{\downarrow})]\\&=P_j^{\uparrow}[(k^2-k_j^{\uparrow}k_j^{\downarrow})/(k^2+k_j^{\uparrow}k_j^{\downarrow})],\qquad j=1,2.\end{aligned} \tag{2.122}$$

Here we use proportional relations in the state density $D_j^{\uparrow,\downarrow}$ and magnitude of the wave vectors $k_j^{\uparrow,\downarrow}$. The modification factor $A_{f/i}=[(k^2-k_j^{\uparrow}k_j^{\downarrow})/(k^2+k_j^{\uparrow}k_j^{\downarrow})]$ depends on the electron penetration to the barrier and on $k_j^{\uparrow}k_j^{\downarrow}$ and $-1<A_{f/i}<1$. Usually, this factor is positive but for sufficiently low barrier becomes small or negative. In such case the electron spins reverse at the interface of ferromagnetic/insulator films.

References

[1] E. Fulcomer and S. H. Charap, *J. Appl. Phys.*, **43**, 4190 (1972)
[2] N. C. Koon, *Phys. Rev. Lett.*,**78**, 4865 (1997)
[3] U. Hartmann (ed) (1999) *Magnetic Multilaters and Giant Magnetoresistance*. Springer Berlin Heidelberg New York
[4] E. H. Sondheimer, *Adv. Phys., suppl. Phil. Mag.,1*. No.1 (1952)
[5] N. H. Mott, *Proc. Roy. Soc.* A**153**, 699 (1936)

[6] I. A. Campbell and A. Fert, *Ferromagnetic Materials*, E. P. Wohlfarth(Ed), North-Holland Pub. Co., Chap. 9, 747 (1982)
[7] J. Barnas, A. Fuss, R. E. Camley, P. Grundberg, and W. Zinn, *Phys. Rev. B*, **42**, 8110 (1990)
[8] H. Sakakima, E. Hirota and Y. Kawawake, *J.MMM*, **184**, 49 (1997)
[9] J. G. Simmons, *J. Appl. Phys.* **34**, 1793 (1963)
[10] J. C. Slonczewski, *Phys. Rev. B*, **39**, 6995 (1989)
[11] M. Julliere, *Phys. Lett.* **54A**, 225 (1975)
[12] S. Maekawa and D. Gafvert, *IEEETrans.MAG,* **MAG-18**, 707 (1982)

3 GMR Devices of Metallic Multilayers

3.1 Exchange Coupled Metallic Multilayers

Since the discovery of giant magnetoresistance (GMR) in Fe/Cr multilayers by Baibich *et al.* [1] and by Binasch *et al.* [2], many studies have been conducted on the phenomenon, which has been attributed to the interlayer antiferromagnetic coupling between the magnetic layers across the nonmagnetic layers [3]. The GMR effect is attractive not only from the point view of basic physics, but also from the applicational point of view, as these metallic mulilayers show much larger magnetoresistance (MR) ratios than conventional magnetic films films, such as NiFe, and the MR sensors using these metallic multilayers were investigated and have come into commerce recently. In this section, the physics of the exchange coupling, the mechanism of the GMR effect, the dependence of the MR properties on the multilayered structure, and the thermal stability of the MR ratios are discussed.

3.1.1 Interlayer Coupling in Metallic Multilayers

The mechanism of the GMR is explained as follows: conduction electrons suffer spin dependent scattering at the interfaces of nonmagnetic and magnetic layers for the antiparallel spin configuration, but not for the parallel spin configuration, as shown in Fig. 3.1. As the result, the resistance of the multilayers is high for the antiparallel spin configuration and low for the parallel spin configuration. The antiparallel spin configuration is realized by interlayer exchange coupling between

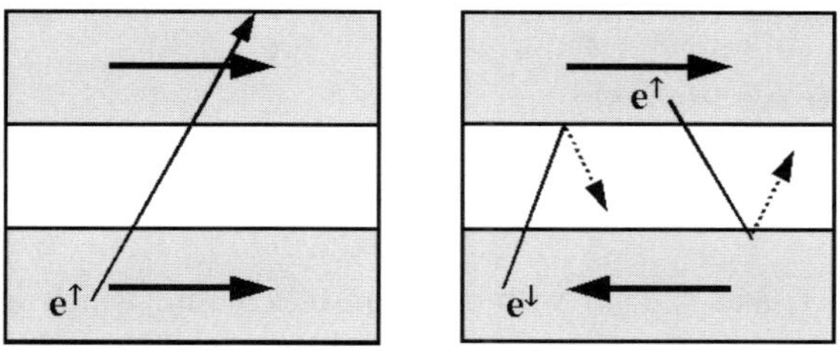

(a) (b)

Fig. 3.1. Schematic illustration of spin dependent electron scattering in a multilayered film for **(a)** parallel spin configuration and **(b)** antiparallel spin configuration

the magnetic layers through the nonmagnetic layers. Parkin *et al.* reported that the magnetoresistance (MR) ratios and the magnitude of the interlayer exchange coupling, J, showed RKKY-like oscillation with nonmagnetic layer thickness in [Co/Ru], [Co/Cr] and [Fe/Cr] multilayers [4] and very large MR ratios (115%) were obtained in [Co/Cu] multilayers at 4.2 K with application of 15 kOe [5].

In the case of RKKY coupling for a point defect, J is given as follows [6]:

$$J \propto \{2k_FR\cos(2k_FR) - \sin(2k_FR)\}/R^4. \tag{3.1}$$

The interlayer exchange coupling between the i-th and j-th layers in multilayers, J_{ij}, is obtained by integrating (3.1) [7] :

$$J_{ij} \propto \sin(2k_FR)/R^2. \tag{3.2}$$

The strength decreases proportionally to the square of the distance between the magnetic layers, changing the sign with an oscillation period of π/k_F.

Figure 3.2 shows the dependence of the MR ratios on the Cu layer thickness, t, in [$Ni_{80}Fe_{15}Co_5$(3)/Cu(t)/Co(3)/Cu(t)] multilayers prepared by sputtering [8], where the value in ( ) indicates the layer thickness in nm. The MR ratios take

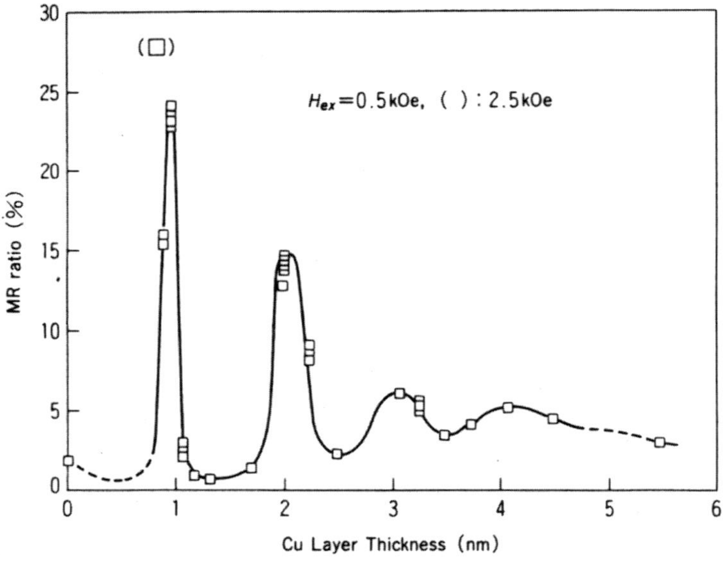

Fig. 3.2. Dependence of MR ratios on the Cu layer thickness in [NiFeCo/Cu/Co/Cu] multilayers

maxima at around t=0.9, 2, 3.1, and 4.2 nm with a period of 1.1 nm, which is smaller than the value expected from π/k_F. The reason will be discussed later. The MR ratios take maxima when the antiferromagnetic interlayer exchange coupling becomes dominant. This can be understood from the magnetization curves of the multilayers shown in Fig. 3.3. The saturation field, H_S, seems to oscillate with t and J is considered to be ferromagnetic for t = 1.4 nm, and to be antiferromagnetic

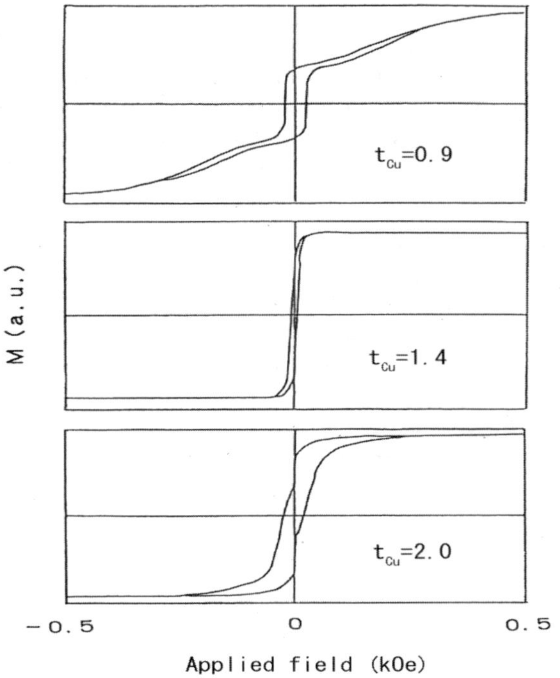

Fig. 3.3. Magnetization curves for [NiFeCo/Cu(t)/Co/Cu(t)] multilayers with t=0.9,1.4, and 2.0nm

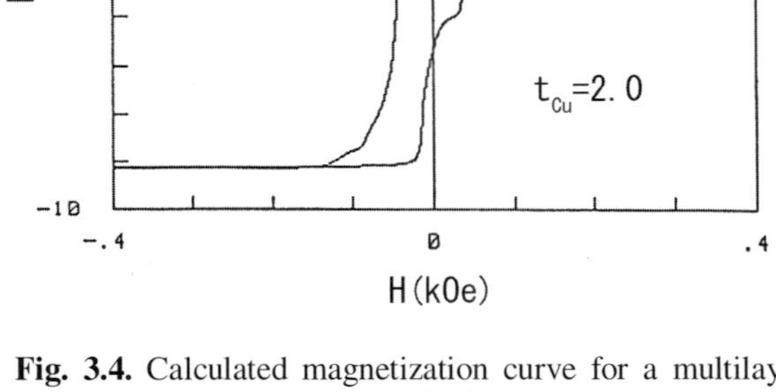

Fig. 3.4. Calculated magnetization curve for a multilayered film composed of magnetic layers having different magnetizations

for t=0.9 and 2.0 where the spin orientations of the magnetic layers are considered to be almost antiparallel at zero field. The remanence observed even for the antiparallel spin configuration is mainly due to the difference in the magnetization between the NiFeCo and Co layers. FMR (ferromagnetic resonance) measurements were also done with [NiFeCo/Cu/Co/Cu] multilayers and RKKY-like oscillations of the resonant fields with the Cu layer thickness were observed [9].

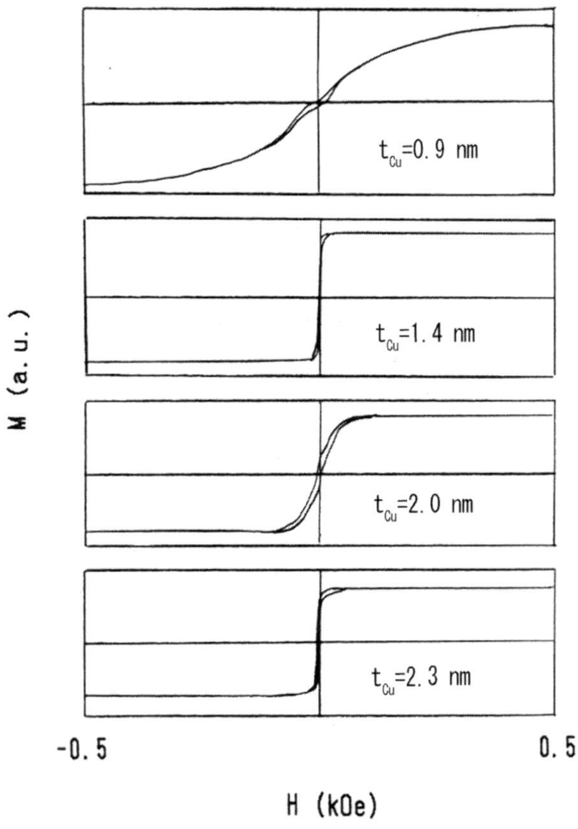

Fig. 3.5. Magnetization curves for [NiFeCo/Cu(t)] multilayers with *t*=0.9, 1.4, 2.0 and 2.3 nm

Magnetization curves of the multilayers can be calculated by minimizing the total energy of the system, *E*, which is given as follows:

$$E=\Sigma\ M_iH_i\cos(\theta_i) - \Sigma\ K_i\cos A_i(\theta_i\text{-}\phi_i) + \Sigma\ J_{ij}\cos(\theta_i - \theta_j), \qquad (3.3)$$

where θ_i is the angle of the direction of the magnetization, M_i, and θ_i and A_i are the angle of the easy axis and the number of symmetry axes of the crystalline anisotropy, K_i, in the i-th magnetic layer. J_{ij} is the exchange coupling between the i-th and j-th magnetic layers across the nonmagnetic layer. Figure 3.4 shows a calculated magnetization curve using (3.3) for multilayers composed of NiFeCo and Co layers antiferromagnetically coupled across Cu layers [10]. The magnetization curve for t=2.0 shown in Fig. 3.3 is qualitatively explained with the calculated one shown in Fig. 3.4.

Other multilayers such as [NiCo/Cu], [NiCo/Cu/Co/Cu], [NiFe/Cu], [Co/Cu], and [NiFeCo/Cu] multilayers [11] also showed RKKY-like oscillation of the MR ratios with Cu layer thickness, t, with the first maximum at around t=0.9 nm and the second maximum at around t=2.0 nm. Figure 3.5 shows the magnetization curves of [$Ni_{80}Fe_{15}Co_5$(3)/Cu(t)] multilayers for t=0.9, 1.4, 2.0 and 2.3 nm. The saturation field, H_S, oscillates with t, and shows maxima at t=0.9 and 2.0 nm where antiferromagnetic coupling is dominant and shows the minimum at t=1.4 and 2.3 nm where ferromagnetic coupling is dominant. In this case, the magnetic layers have the same magnetization, so that the remanence is almost zero for the multilayers with t=0.9 and 2.0 nm.

3.1.2 MR Properties of Metallic Multilayers

Metallic multilayers composed of magnetic layers, antiferromagnetically exchange coupled across the nonmagnetic layers, show the GMR effect. The MR ratio, $\Delta R/R$, is defined in this book as

$$\Delta R/R=(R_{max}-R_{min})/R_{min} \times 100\ (\%) \tag{3.4}$$

where R_{max} and R_{min} are the maximum and minimum value of the resistance, respectively. The resistance becomes a maximum when the magnetic layers take an antiparallel spin configuration due to the antiferromagnetic exchange coupling near zero field and becomes a minimum when they take a parallel spin configuration with application of the magnetic field, H, larger than the saturation field, H_S, as shown in Fig. 3.6. When the crystalline anisotropy energy, K, is negligibly small, H_S is related to J [12],

$$H_S \propto J/(M_S D_m) \tag{3.5}$$

where D_m and M_S are the thickness and magnetization of the magnetic layers. Therefore, H_S is proportional to J and shows RKKY-like oscillations as shown in Figs. 3.3 and 3.5.

MR curves of the multilayers can also be calculated using (3.3). Phase diagrams and MR curves were studied for the multilayers composed of two magnetic layers antiferromagnetically coupled across a non magnetic layer by Folkerts [13]. Now let us consider a simple case, where the magnetization, M, and the thickness of the two magnetic layers are equal and an external magnetic field, H, is applied in the film plane, then the total energy of the system is given by

$$E= - (1/2)HM(\cos\phi_1+\cos\phi_2) - J_{AF}\cos(\phi_2-\phi_1) + E_K \tag{3.6}$$

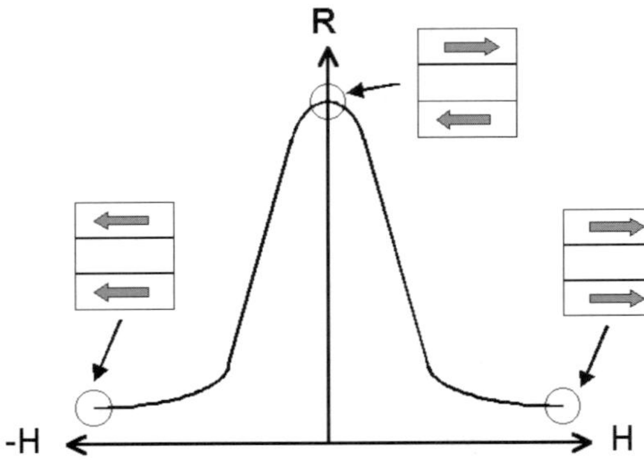

Fig. 3.6. Spin configurations and MR curve for the exchange coupled multilayered film

where ϕ_1 and ϕ_2 are the angle between H and M, and E_K is the anisotropy energy. In the case of uniaxial in-plane anisotropy, E_K is given as

$$E_K=(1/2)K_U(\cos^2\phi_1+\cos^2\phi_2) \qquad \text{for } H \perp K_U, \tag{3.7}$$

and

$$E_K=(1/2)K_U(\sin^2\phi_1+\sin^2\phi_2) \qquad \text{for} H \,//\, K_U. \tag{3.8}$$

The stable configurations are obtained by minimizing E with $\partial E/\partial\phi_1=0$ and $\partial E/\partial\phi_2=0$. The magnetic phases and the energy for $H \perp K_U$ [13, 14] are:
(a) $\phi_1=\phi_2=0$:

$$E=K_U(1-2h+p) \tag{3.9}$$

(b) $\phi_1=-\phi_2$ with $\cos\phi_1=h/(2p+1)$:

$$E=K_U(-h\cos\phi_1-p) \tag{3.10}$$

(c) $\phi_1=\phi_2=\pi$:

$$E=K_U(1+2h+p) \tag{3.11}$$

where $h=HM/2K_U$ and $p=-J_{AF}/Ku$. Figure 3.7 shows the magnetic phase diagram for this configuration. Transitions between the phases are simple in this case and (a)↔(b) and (b)↔(c) occur at $h=2p+1$ and $h=-(2p+1)$, respectively. Therefore, the saturation field, H_S, is given,

$$H_S=-4J_{AF}/M+H_K, \quad (H_K=2K_U/M) \tag{3.12}$$

in this case. If K_U is negligibly small, Eq.(3.10) becomes equal to Eq.(3.5).

The magnetic phases and the energy for $H \,//\, K_U$ [13, 14] are:
(a) $\phi_1=\phi_2=0$:

$$E=K_U(-2h+p) \tag{3.13}$$

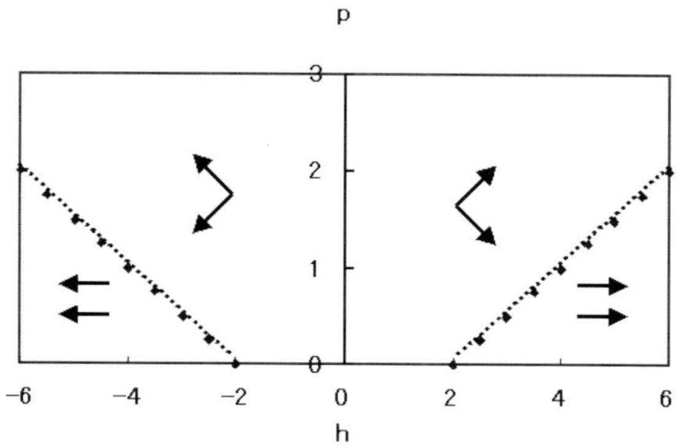

Fig. 3.7. Magnetic phase diagram for the $H \perp K_U$ configuration

(b) ϕ_1=-ϕ_2 with $\cos\phi_1$=h/(2p-1)>0:

$$E=K_U(1-p-h\cos\phi_1) \tag{3.14}$$

(c) ϕ_1=0, ϕ_2=π:

$$E= K_U(-p) \tag{3.15}$$

(d) ϕ_1=-ϕ_2 with $\cos\phi_1$=h/(2p-1)<0:

$$E=K_U(1-p-h\cos\phi_1) \tag{3.16}$$

(e) ϕ_1=ϕ_2=π:

$$E=K_U(2h+p) \tag{3.17}$$

Transitions between these phases are not simple in this case [13] and the GMR film with this configuration sometimes shows a peculiar MR curve as shown in Fig. 3.8. The MR curve was compared with the calculated one using (3.11) and qualitative agreement was obtained [14].

Generally speaking, H_s of the GMR film is large so long as K is not small, as H_s has relation with J and K shown in (3.10). For example, H_s is large (3~10 kOe) for [Co/Cu] multilayers [5,15], as the K of the Co layer is not small. It is desirable from the applied point of view that the mulatilayers have small H_s and show a low-field GMR. For this purpose, miltilayers composed of magnetic layers having small K and small magnetostriction, λ, have been studied [8, 9]. It is found that multilayers having Ni-rich NiFeCo layers, such as [NiFeCo/Cu/Co/Cu] [8,10] and [NiFeCo/Cu] [8,11], show low-field GMR. It is found that [M(=Co,Fe,Ni)/Cu(t)] multilayers show RKKY-like oscillations with Cu layer thickness and show firstand second MR peaks at around t=0.9 and 2.0 nm, respectively. Figure 3.9 shows MR curves for [NiFeCo(3)/Cu(t)/Co(3)/Cu(t)] with t=0.9 and t=2.0, which correspond to the first and second MR peak shown in Fig. 3.2. The MR ratio for the first peak is larger than that of the second peak, but the H_s of the first peak is much larger than that of the second peak, as J is given by (3.3). So, it is desirable to use the second MR peaks for practical applications that need low-field GMR.

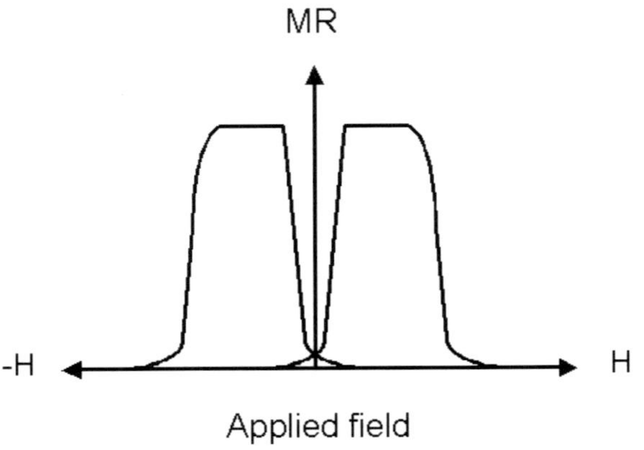

Fig. 3.8. Example of MR curves with the H // K_U configuration

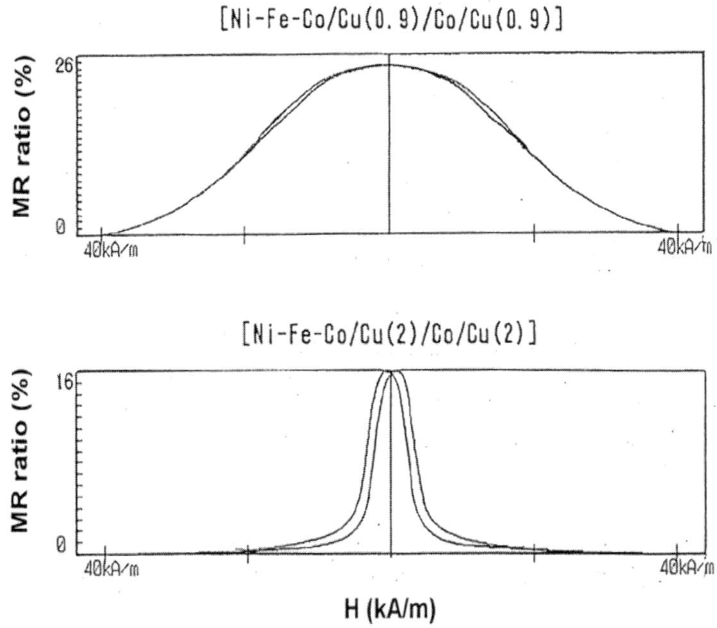

Fig. 3.9. MR curves for [NiFeCo/Cu(t)/Co/Cu(t)] multilayers with t=0.9 and 2.0 nm

The dependence of the MR ratios on the composition of the magnetic layers was studied experimentally and theoretically by various researchers. The results are summarized by Miyazaki *et al.* [16] as shown in Fig. 3.10, where the experimental results of the MR ratios for the first MR peaks of the [Fe-Co-Ni/Cu] multilayers are plotted with a solid line as a guide for the eyes and the calculated result by Inoue e*t al.* [17] is shown with a dashed line. The MR ratios of

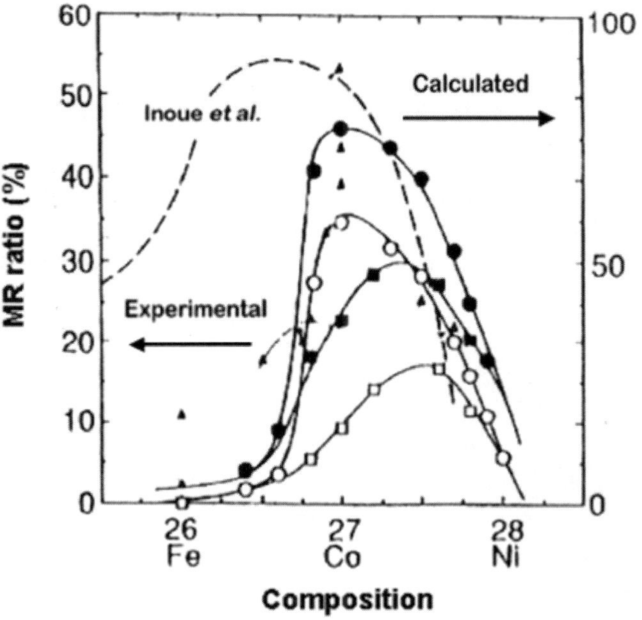

Fig. 3.10. Composition dependence of MR ratios for [Fe-Co-Ni/Cu] multilayers [16]

experimental results take the maximum at around an electron number of 27, which qualitatively agrees with the theoretical result, though the theoretical calculation takes only the interfacial scattering into consideration. In reality, the MR ratio depends upon the interfacial and bulk scattering, as well as the layer thickness and the mean free path of conduction electrons. But it may be said that the spin-dependent scattering that contributes to the GMR effect is much larger at the Co/Cu interfaces than at the Ni/Cu interfaces. This may be understood from the experimental result that an insertion of thin Co layers between the Ni(Fe) and Cu layers increases the MR ratios (see Fig. 3.24). The spin asymmetry factor for bulk scattering in the Co layer is considered to be larger than that in the Ni layer. Generally speaking, multilayers composed of Co-rich magnetic layers show larger MR ratios than those composed of Ni-rich magnetic layers, so long as the layered structure is the same.

3.1.3 Dependence of the MR Properties on the Film Structures

Metallic multilayers are prepared by sputtering or by evaporation. It is very important that the multilayers have a clear and continuous layered structure with smooth interfaces to show the GMR effect owing to magnetic scattering at the interfaces between the magnetic and nonmagnetic layers as shown in Fig. 3.1. Generally speaking, the sputtering method suits the preparation of a flat and continuous layered structure better than the evaporation method, as the sputtered atoms have sufficient energy for migration, resulting in the growth of a layered structure, while the evaporated atoms have small energy, resulting in the growth of an island structure. Figure 3.11 shows a cross-sectional TEM image for the [NiFeCo/Cu(0.9)/Co/Cu(0.9)] multilayered film prepared by sputtering, which shows the first MR peak as shown in Figs. 3.2 and 3.9. It is found that the film has a clear and continuous layered structure, even though the Cu layer is extremely thin, such as 0.9 nm. Figure 3.12 shows a small-angle XRD pattern for the film with calculated satellite peak positions denoted by arrows. The artificial periodicity of the layered structure is confirmed by the fact that the clear satellite peaks are found at the positions calculated with the step model [18] using the designed structure.

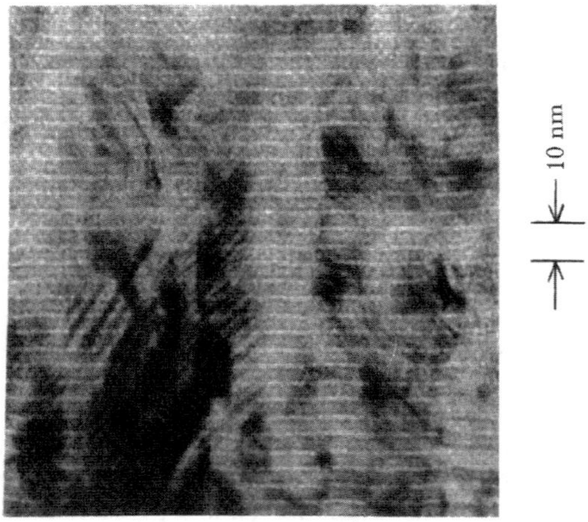

Fig. 3.11. Cross-sectional TEM image for [NiFeCo/Cu(0.9)/Co/Cu(0.9)]

The multilayers prepared by sputtering show clear layered structures. But different structures were observed for evaporated ones. Figure 3.13 shows a cross- sectional TEM image for [Co/Cu] multilayers prepared by evaporation. The evaporated [Co/Cu] multilayered film has a stacking structure of wavy layers, caused by the island growth. Figure 3.14 shows the dependence of MR ratios on the Cu layer thickness for the evaporated and sputtered [Co/Cu] multilayers. Both films showed the second peak of GMR, but only the sputtered one showed the first MR peak at around 0.9 nm. This result may be attributed to the difference in the layered structure between the sputtered and evaporated films, because the wavy structure of the evaporated film causes so-called "orange peel" coupling [19] between the magnetic layers across the nonmagnetic spacer. This Neel-type ferromagnetic coupling becomes dominant for a thinner spacer, such as a 0.9 nm thick Cu layer for the first MR peak. Figure 3.15 shows schematic illustrations of the structures of those films. The incident energy of the sputtering is high and contributes to the migration of the atoms at the film surfaces resulting in the growth of a layered structure, while that of evaporation is low resulting in the growth of island structures. But the interdiffusion at the interfaces is considered to be smaller in the evaporated films than in the sputtered films due to the difference in the incident energy.

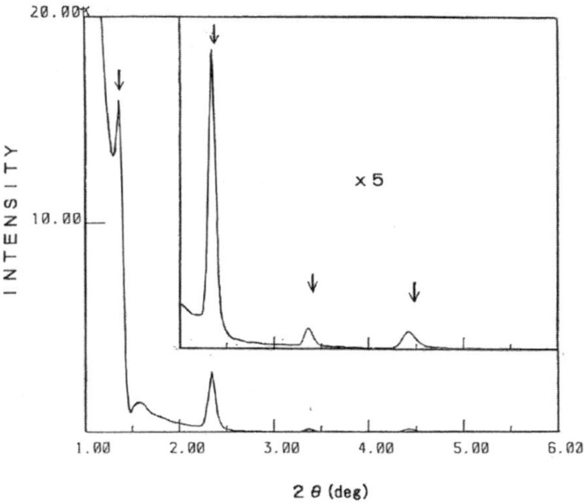

Fig. 3.12. Small-angle XRD patterns for [NiFeCo/Cu(0.9)/Co/Cu(0.9)]

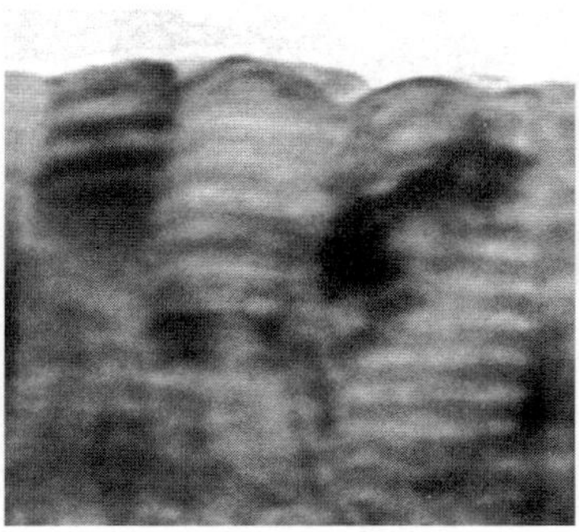

Fig. 3.13. Cross-sectional TEM image for [Co(2)/Cu(2.3)] prepared by evaporation

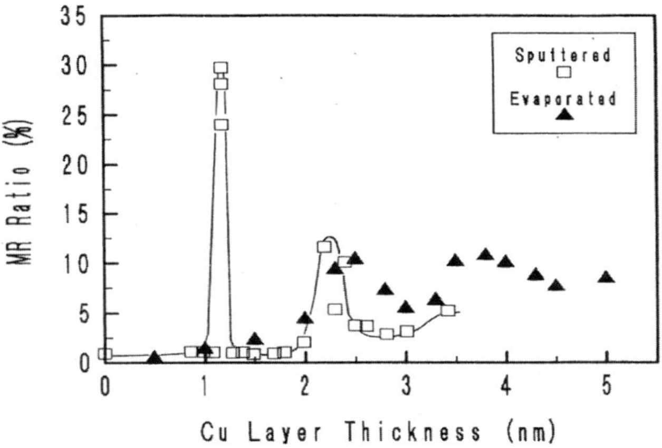

Fig. 3.14. Dependence of MR ratios on Cu layer thickness for [Co/Cu] multilayers

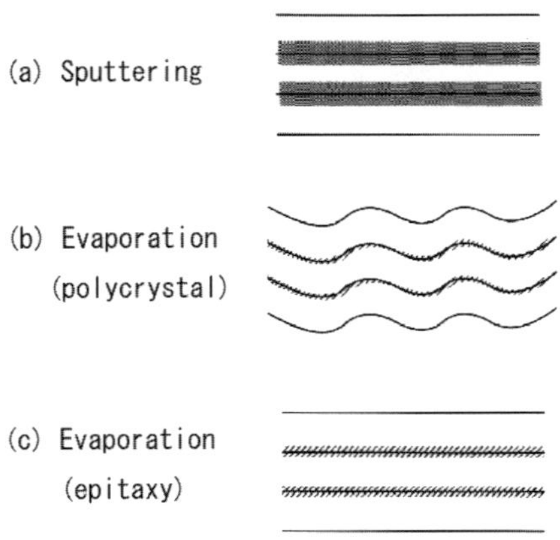

Fig. 3.15. Schematic illustration of the film structures: **(a)** sputtered film, **(b)** evaporated film and **(c)** epitaxially grown film

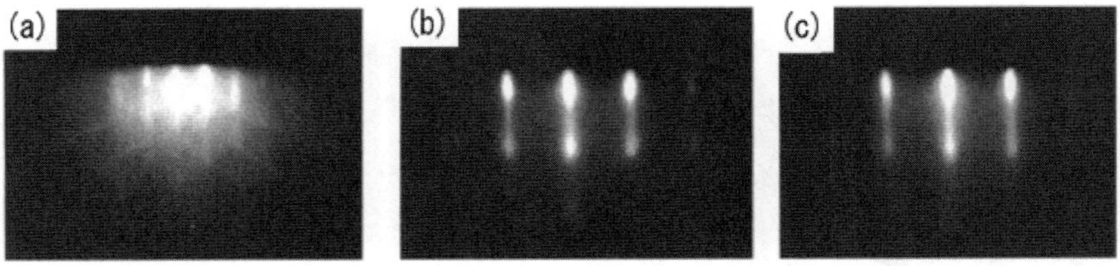

Fig. 3.16. RHEED patterns for the film prepared on Si(100): **(a)** Si(100) along <011> azimuth, **(b)** Si/Cu(100 nm), **(c)** Si/Cu/[Cu(1 nm)/NiFe(3 nm)][20]

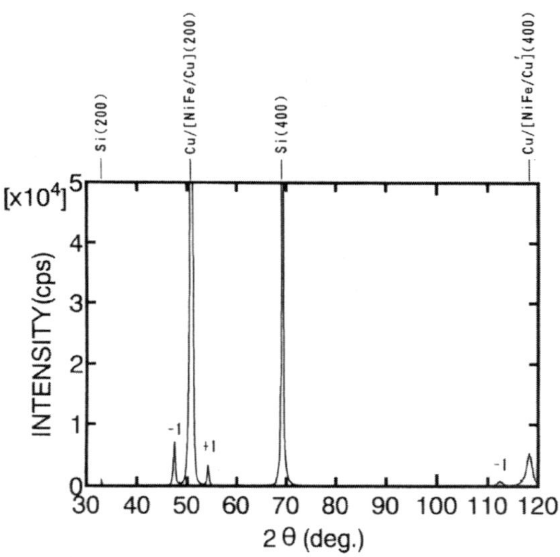

Fig. 3.17. X-ray diffraction patterns for Si(100)/Cu/[NiFe/Cu] epitaxial film

Epitaxial film growth using MBE may enable one to form an ideal layered structure with small interdiffusion as the incident energy is small and epitaxial growth is effective to suppress the island growth as shown in Fig. 3.15. Epitaxial growth of [NiFe/Cu] multilayers on Si(100)/Cu and Si(111)/Ag/Cu were studied using MBE [20], where Cu and Ag/Cu were used for the buffer layers. Figure 3.16

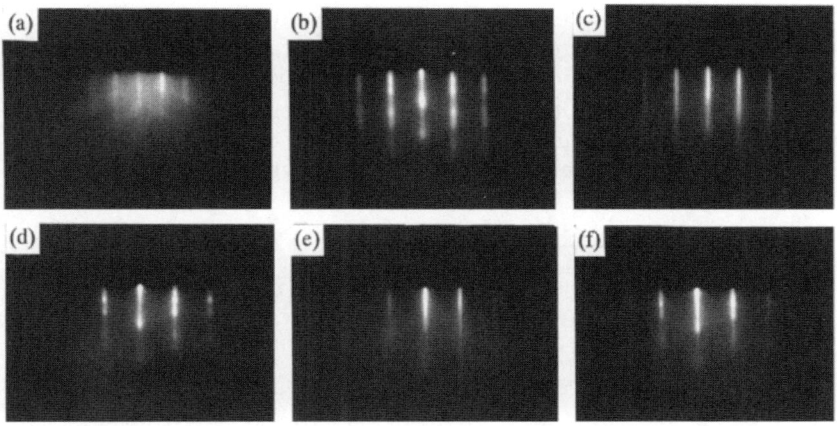

Fig. 3.18. RHEED patterns for films prepared on Si(111): **(a)** Si(111) along <111> azimuth, **(b)** Si/Ag(7.6 nm), **(c)** after annealing at 100°C, **(d)** Si/Ag/Cu(50 nm), **(e)** after annealing at 200°C, **(f)** Si/Ag/Cu/[NiFe(3 nm)/Cu(0.9 nm)][20]

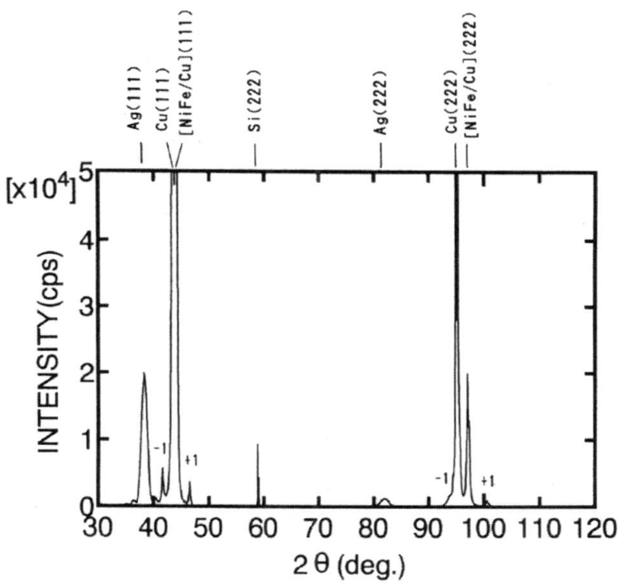

Fig. 3.19. X-ray diffraction patterns for Si(111)/Ag/Cu/[NiFe/Cu] epitaxial film

shows RHEED patterns of the films grown on Si(100) substrates, where (a) is the pattern of the Si(100) surface along the <110> azimuth, (b) is the pattern after the deposition of 100 nm thick Cu buffer layer and (c) is the pattern after the preparation of [Cu(1nm)/NiFe(3nm)]^{20} multilayers. Figure 3.17 shows the X-ray diffraction pattern of the multilayered film prepared on Si(100). Figure 3.18 shows RHEED patterns for the films grown on Si(111) substrates, where (a) is the pattern of the Si(111) along the <111> azimuth, (b) is the pattern after the deposition of a 7.6 nm thick Ag buffer layer, (c) is the pattern after annealing at 100 °C, (d) is the pattern after the deposition of a 50 nm thick Cu buffer layer, (e) is the pattern after annealing at 200° C, (f) is the pattern after the preparation of [NiFe(3nm)/ Cu(0.9nm)]^{20} multilayers. Figure 3.19 shows the X-ray diffraction pattern of the multilayered film prepared on Si(111). Epitaxial film growth is confirmed from

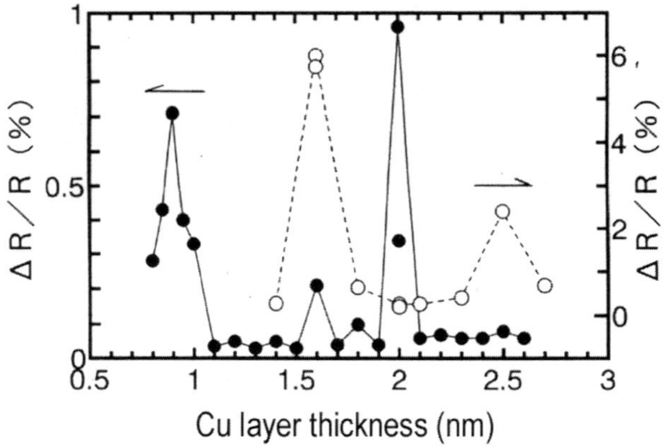

Fig. 3.20. Dependence of MR ratio on the Cu layer thickness, t, for Si(100)/Cu(100nm)/ [NiFe(3nm)/Cu(t)]^{20}: ●, and Si(100)/Cu(50nm)/ [NiFe(1nm)/Cu(t)]^{50}: ○

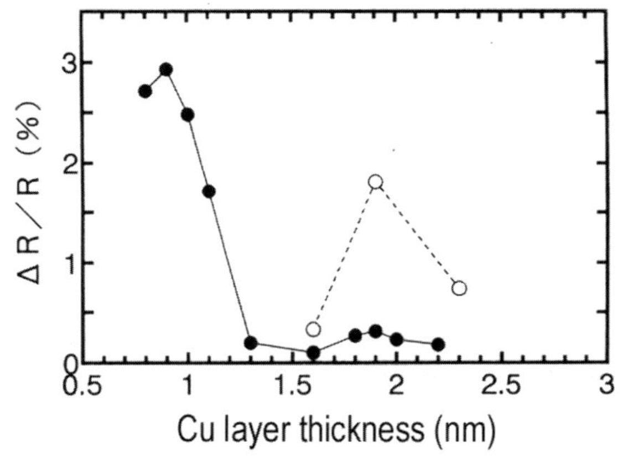

Fig. 3.21. Dependence of MR ratio on the Cu layer thickness, t, for Si(111)/Ag(7.6nm)/ Cu(50nm)/[NiFe(d)/Cu(t)]^{N}, where N=20 and d=3 for ●, and N=50 and d=1 for ○

the RHEED patterns. Streak patterns are observed for both cases and the formation of flat surfaces is confirmed in these epitaxial films. Clear (100) and (111) orientations are observed by the X-ray diffraction patterns for [NiFe/Cu] multilayers prepared on Si(100) and Si(111), respectively. The orientations in the stacking direction are Si(100)/Cu(100)/[NiFe/Cu](100) and Si(111)/Ag(111)/Cu(111)/[NiFe/Cu](111) for the films prepared on Si(100) and Si(111), respectively. The orientations in the film plane are Si<011>/Cu<001>/[NiFe/Cu]<001> and Si<011>/Ag<011>/Cu<011>/[NiFe/Cu]<011> for the films prepared on Si(100) and Si(111), respectively [20].

The dependence of the MR properties on the crystal orientation was studied using these epitaxial films. Figures 3.20 and 3.21 show the dependence of the MR ratios on the Cu layer thickness, *t*, for the (100) oriented and the (111) oriented [NiFe/Cu] epitaxial films, respectively. RKKY-like oscillations are observed for both films. The (100) oriented film shows a long (~1.1 nm) and a short (0.4~0.5 nm) oscillation period, while the (111) oriented one shows only one long (~1.1nm) oscillation period. These results agree with the theory developed by Bruno *et al.* [21], which predicts that the oscillation periods, λ, are obtained from the relevant caliper, q, in the Fermi surface of the nonmagnetic layers. Figure 3.22 shows a

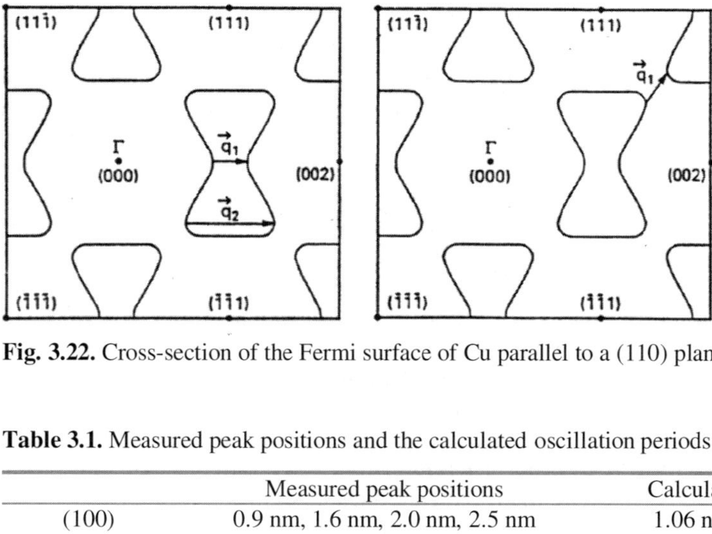

Fig. 3.22. Cross-section of the Fermi surface of Cu parallel to a (110) plane

Table 3.1. Measured peak positions and the calculated oscillation periods for a Cu spacer

	Measured peak positions	Calculated periods
(100)	0.9 nm, 1.6 nm, 2.0 nm, 2.5 nm	1.06 nm, 0.46 nm
(111)	0.9 nm, 1.9 nm	0.94 nm

Table 3.2. H_s and J for epitaxial $[NiFe(3)/Cu(0.9)]^{20}$ film

	H_S (kOe)	J (erg/cm^2)
(100)	~ 0.3	~ 1.8×10^{-2}
(111)	~ 2.5	~ 1.5×10^{-1}

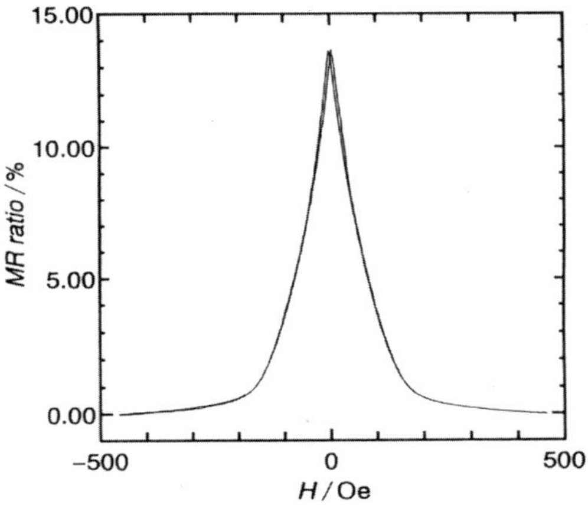

Fig. 3.23. MR curve for Si(111)/Cu(5nm)[NiFe(1)/Cu(1.6nm)]

cross-section of the Fermi surface of Cu parallel to the (110) plane. If the stacking direction is normal to (100), there exist q_1 and q_2 in the "dog's bone orbit" as shown in Fig. 3.22 and the two periods are obtained with $\lambda_1=2\pi/q_1$ and $\lambda_2=2\pi/q_2$. If the stacking direction is normal to (111), there exists only one q_1 as shown in Fig. 3.22 and only one period is obtained with $\lambda_1=2\pi/q_1$. The results of the theoretical calculation for the oscillation periods [22] are also shown in Table 3.1 for comparison. The saturation field, H_S, and the magnitude of the interlayer exchange coupling, J, for the first MR peak of (100) and (111) oriented epitaxial $[NiFe(3)/Cu(0.9)]^{20}$ are summarized in Table. 3.2. The values of H_S and J for the (100) oriented evaporated epitaxial film were found to be much smaller than those for the (111) oriented one and the values were almost the same as those for the sputtered multilayered film. The new MR peaks found at around t=1.6 and 2.5 nm for the (100) oriented epitaxial film, which was not observed in sputtered films, are considered to correspond to the short period shown in Table 3.1. The MR ratios shown in Fig. 3.20 are rather small, but this is due to the shunting effect of the 100 nm thick Cu buffer layer. Larger MR ratios are obtained by reducing the buffer layer thickness as shown in Fig. 3.23. The MR ratios obtained for the epitaxial [NiFe/Cu] film having a thin buffer layer are considered to be larger than those of sputtered [NiFe/Cu] films. Therefore, epitaxial multilayers are considered to have better layered structure than sputtered ones as shown in Fig. 3.15.

3.1.4 Thermal Stability of the MR Ratios

It is important to study the thermal stability of the MR ratios of GMR films for practical applications of these films to MR sensors. Multilayers composed of [NiFeCo(3)/Cu(t)] and [NiFeCo(3)/Cu(t)/Co(3)/Cu(t)] were prepared by sputtering. The Cu layer thickness, t, was fixed as 0.9 nm for the first MR peak and 2.0 nm for the second MR peak. The films were annealed in a vacuum and the MR properties were measured at RT. Figures 3.24 and 3.25 show the annealing temperature dependence of MR ratios measured at RT by applying an external

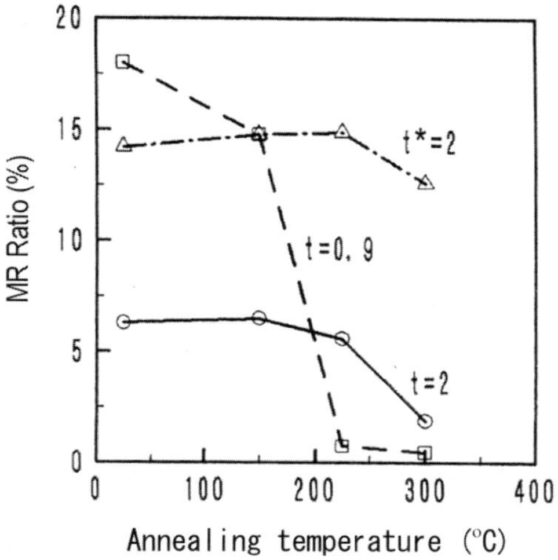

Fig. 3.24. Annealing temperature dependence of MR ratios for [NiFeCo/Cu(t)][40] and [Co(0.4)/NiFeCo/Co(0.4)/Cu(t*)][40] multilayers

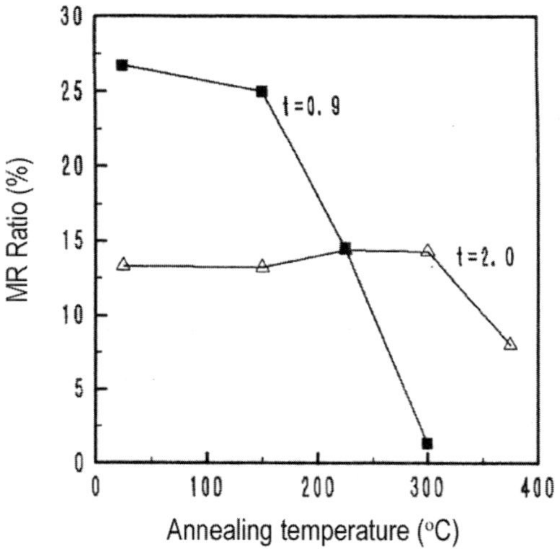

Fig. 3.25. Annealing temperature dependence of MR ratios for [NiFeCo/Cu(t)/Co/Cu(t)][20] multilayers

magnetic field of 2.5 kOe. The films with the Cu layer thickness, t=2.0 nm, showed better thermal stabilities of the MR ratios than those with t=0.9 nm. Small-angle X-ray diffraction measurements were also carried out to study the structure changes of these multilayers. Decreases of the diffraction peak intensities corresponding to the artificial periodicities and the peak shifts towards the high angle side were observed by annealing above 225 °C for all multilayers. Therefore, annealing above 225 °C causes degradation and shrinkage of the artificial periodicity, which was estimated as about 3% [23]. Deterioration of the MR ratios shown in Figs. 3.24 and 3.25 may be explained with these structure changes for the films with t=0.9 nm, but not for the films with t=2.0 nm. Therefore, the GMR effect is more sensitive to the structural change in films with t=0.9 nm than in films with t=2.0 nm.

The MR ratios of the films having Co layers such as [NiFeCo/Cu(t)/Co/Cu(t)] are thermally more stable than that of the films without Co layers such as [NiFeCo/Cu(t)] when the Cu layer thickness, t, is the same. Insertion of thin Co layers between the NiFeCo and Cu layers is effective to increase the MR ratios and the thermal stability as shown in Fig. 3.24. These results imply that the Co/Cu interfaces are more thermally stable than the Ni/Cu interfaces. The solubility of Cu in Co is lower than in Ni. This may contribute to the difference in the thermal stability of the MR properties between the multilayers having Co/Cu interfaces and Ni/Cu interfaces. Actually, the MR ratios of [Co/Cu] sputtered multilayers were thermally stable up to 375 °C [24] as shown in Fig. 3.26. The MR ratio of the sputtered [Co/Cu] increased by annealing, but that of evaporated [Co/Cu] did

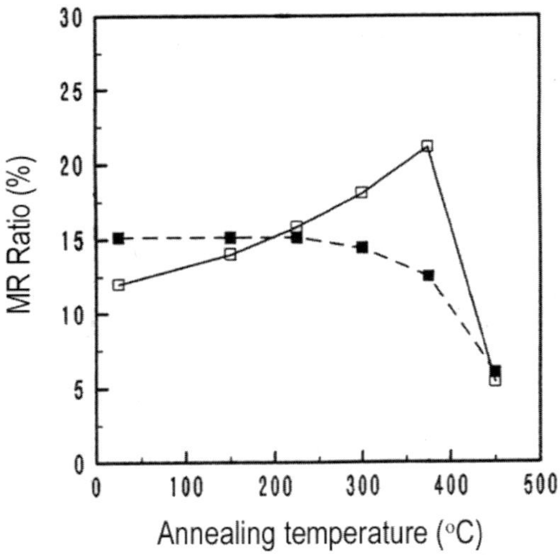

Fig. 3.26. Annealing temperature dependence of MR ratios for [Co/Cu] multilayers: **(a)** □ sputtered and **(b)** ■ evaporated

not show such a tendency. The interfaces of the sputtered [Co/Cu] film are considered to have an interdiffusion region as shown in Fig. 3.15, when the film is prepared. Then annealing below 375 °C probably caused segregation of Cu from Co at the interfaces, as the solubility of Cu in Co is low, reducing the interdiffusion region, and results in an improvement of the interface quality. But in the case of evaporated [Co/Cu] film, the interdiffusion region is considered to be small as shown in Fig. 3.15. As a result, improvement of the interface quality by annealing is considered to be small for the evaporated case. The reason why the MR ratios of the evaporated films are smaller than those of the sputtered films is probably due to the interface roughness of the evaporated films caused by the island growth as shown in Fig. 3.15. But larger MR ratios are obtained for the evaporated films epitaxially grown (by MBE), having good interfaces, as mentioned in Sect. 3.1.3.

3.2 Magnetic Sensors

GMR films composed of [NiFeCo/Cu] or [NiFeCo/Cu/Co/Cu] multilayers show considerably large MR ratios with application of small magnetic fields as mentioned in Sect. 3.1.2. Figure 3.27 shows MR curves for the first (t=0.9 nm) and second (t=2.0 nm) MR peak of [NiFeCo/Cu] multilayers. The first peak shows a larger MR ratio than the second peak, but the saturation field, H_s, is larger than the second peak and the thermal stability is not so high as the second peak, as mentioned in Sect. 3.1.4. Therefore, films with second MR peak are used for practical applications. NiFe films are used for conventional MR sensors. But the MR ratio is about 2~2.5%. Therefore, MR sensors using GMR films that show larger MR ratios, such as 10%, are expected to show higher output than conventional ones. Another merit of GMR films is that the films show isotropic MR properties unlike the conventional single layered films, such as NiFe, which show anisotropic MR properties and are called AMR (anisotropic MR) films.

3.2.1 Static Characteristics

MR sensors are fabricated by patterning the GMR or AMR films into a zigzag shape as shown in Fig. 3.28, where L and W are the length and width of the pattern. It is desirable to increase the aspect ratio, L/W, to obtain higher output. However, AMR films do not show any MR effect, when the current, I, flows parallel to the magnetic field (shown as $H_{//}$ in the figure). Therefore, the MR sensors using AMR films are designed so that the current direction is perpendicular to the magnetic field (shown as $H_{\perp}$ in the figure) to be sensed. But this configuration degrades the sensitivity of the AMR sensors, as the demagnetization field is increased with increase of the aspect ratio, L/W. As a result, it was difficult for AMR sensors to realize compatibility of high output and high sensitivity. GMR films offer compatibility, as the films show isotropic MR properties. It is possible to design GMR sensors so that the zigzag pattern has high aspect ratio and the current direction is parallel to the magnetic field to be sensed. In this configuration, it is possible to increase the out- put and the sensitivity of GMR sensors by increasing the aspect ratio, L/W.

MR sensors are fabricated [25] by patterning GMR films composed of $[NiFeCo/Cu(2nm)]^{30}$ multilayers as shown in Fig. 3.27 and NiFe films into zigzag shapes. First, the static characteristics of the MR sensors are studied. Figure 3.29 shows the MR characteristics of fabricated AMR and GMR sensors for I // H and $I \perp H$ configurations. It is found that the GMR sensor with the I // H configuration shows higher output and higher sensitivity than the AMR sensor with the $I \perp H$

configuration. No output is obtained for the AMR sensor with the I // H configuration. The outputs for the I // H and $I \perp H$ configurations are almost the same in the GMR sensor, but the sensitivity for I // H is higher than that for $I \perp H$ due to the difference in the demagnetization field.

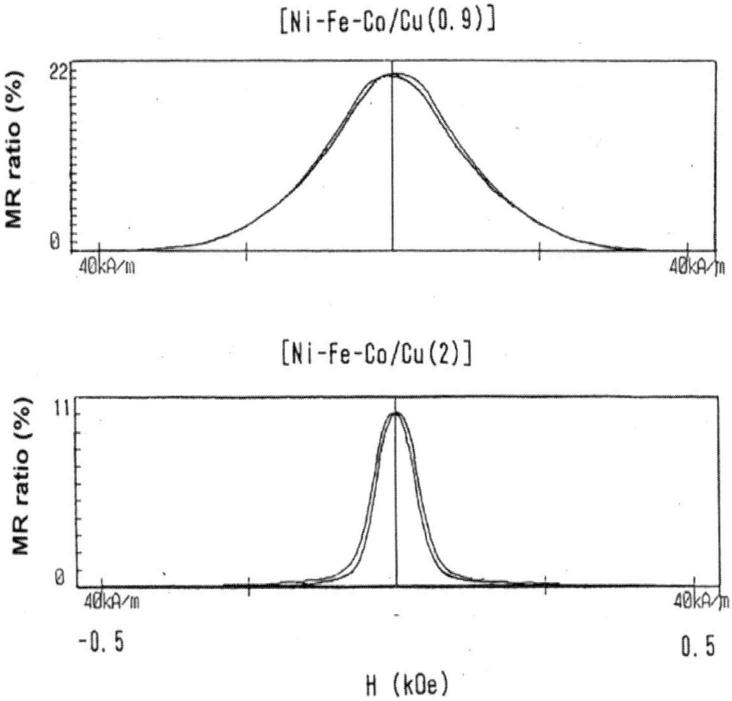

Fig. 3.27. MR curves for [NiFeCo/Cu(t)] with t=0.9 and 2.0 nm

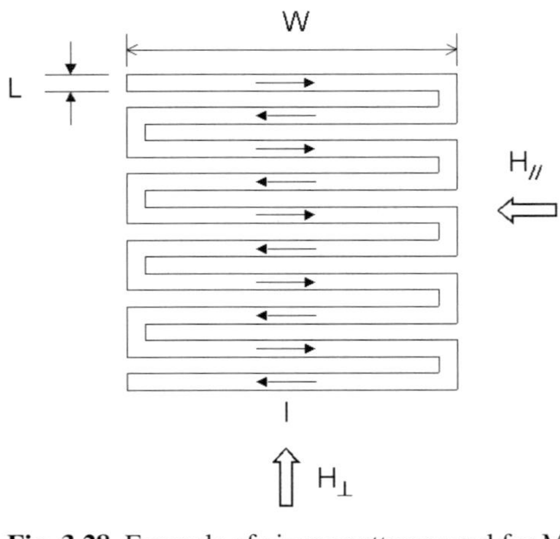

Fig. 3.28. Example of zigzag patterns used for MR sensors

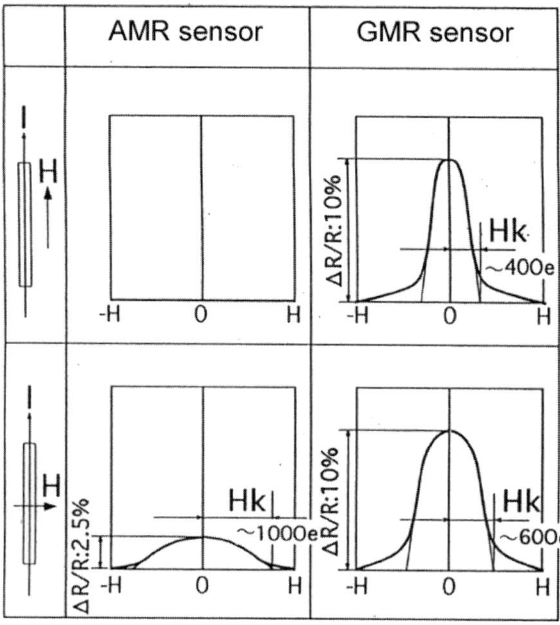

Fig. 3.29. MR characteristics of AMR and GMR sensors for I // H and I ⊥ H configurations

3.2.2 Dynamic Characteristics

The dynamic characteristics of fabricated GMR sensors were studied with a capstan motor having a magnetized rotor with a pitch of 300 μm, as shown in Fig. 3.30. This type of capstan motor is used for VCR and the MR sensor is used for the rotation control. The MR sensor is required to show high output and high sensitivity to cope with the capstan motor with a narrower pitch, as the diameters of the rotor are becoming smaller with the downsizing of VCRs. The MR sensor is mounted on the substrate facing the rotor with an air gap. The mounting of a conventional MR sensor requires manual adjustment, as the signal output strongly depends upon the gap length and the allowance is small.

Figure 3.31 shows the dependence of the output voltage on the air gap length for GMR and AMR sensors. The GMR sensor shows 2~3 times higher output voltage than the conventional AMR sensor for whole range of the gap length. The GMR sensor realizes a 2 times wider range of the gap length where the output voltage is higher than 40 mV, compared with the AMR sensor. This increased allowance for the gap length enables the mounting of the GMR sensor with a robot. Figure 3.32 shows the dependence of the amplitude of modulation noise on the air gap length. The modulation noise for the GMR sensor keeps a small amplitude for a wide range of air gap length, while that of the AMR sensor increases with increase of the air gap length.

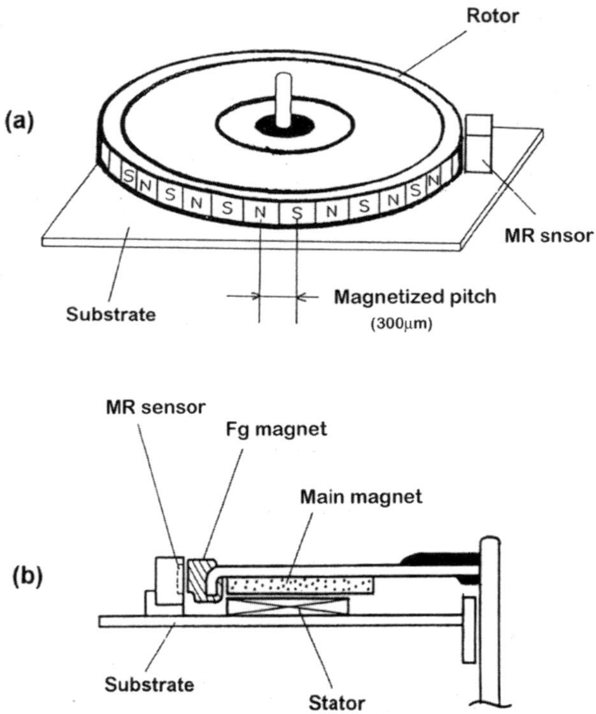

Fig. 3.30. Illustration of a capstan motor with an MR sensor: **(a)** general view and **(b)** cross-section

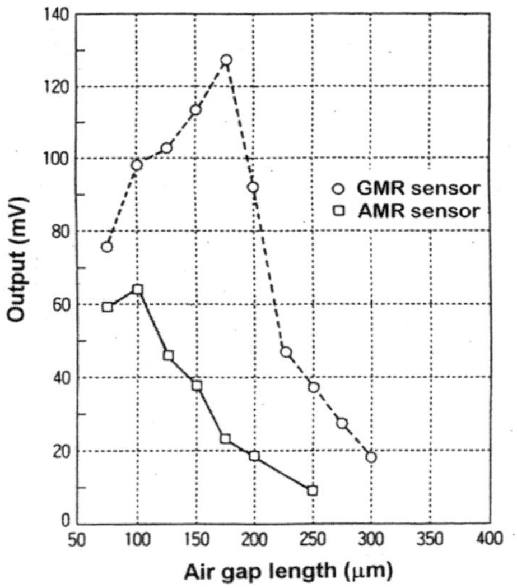

Fig. 3.31. Dependence of the output voltage on the air gap length

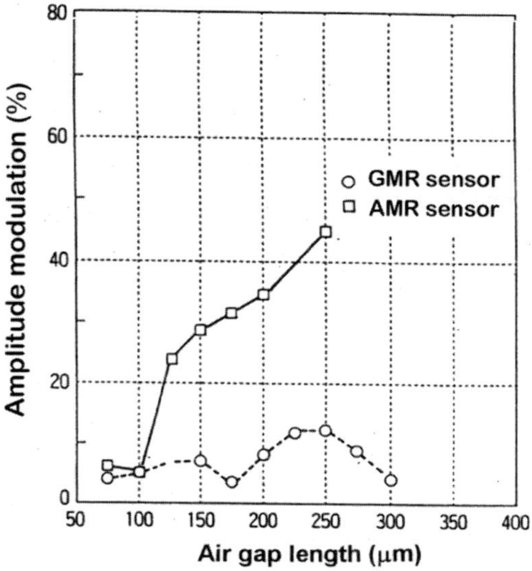

Fig. 3.32. Dependence of the degree of amplitude modulation on the air gap length

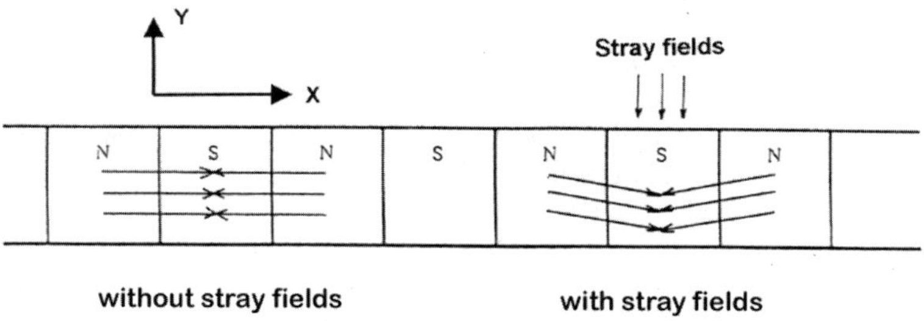

Fig. 3.33. Schematic view of the magnetic flux direction near the MR sensor

The modulation noise is due to the stray fields from the main magnet, as the direction of the leakage flux from the main magnet is almost orthogonal to that of the signal flux from the rotor as shown in Fig. 3.33. The AMR sensor only detects the X component of the magnetic field, $\boldsymbol{H}$, and the output is proportional to $|\boldsymbol{H}|X/(X^2+Y^2)^{1/2}$,while the GMR sensor detects the whole strength of the magnetic field, $|\boldsymbol{H}|$, as the GMR effect is isotropic. This is the reason why the modulation noise caused by the main magnet is observed in the AMR sensor but not in the GMR sensor. The distance between the MR sensor and the magnetized rotor, D_S, is smaller than that between the MR sensor and the main magnet, D_N, when the air gap length is small but the difference between D_S and D_N becomes smaller when the air gap length becomes larger. This is the reason why the modulation noise increases with increase of the air gap length, as the S/N ratio is considered to be a function of D_S/D_N, for the AMR sensor. Thus, GMR sensors are found to show higher sensitivities, larger outputs and smaller noise than conventional AMR sensors.

References

[1] M. N. Baibich, J. M. Broto, A. Fert, F. Nguyen Van Dau, F. Petroff, P. Etienne, G. Creuzet, A. Friedrich and J. Chazelas : *Phys. Rev. Lett.* **61**, 2472 (1988).

[2] G. Binasch, P. Grunberg, F. Saurenbach and W. Zinn : *Phys. Rev.* **B39**, 4828 (1989).

[3] N. Hosoito, S. Araki, K. Mibu and T. Shinjo : *J. Phys. Soc. Jpn.* **59**, 1925 (1989).

[4] S. S. P. Parkin, N. More, and K. P. Roche : *Phys. Rev. Lett.* **64**, 2304 (1990).

[5] S. S. P. Parkin, Z. G. Li, and D. J. Smith : *Appl. Phys. Lett.* **58**, 2710 (1991).

[6] M. A. Rudermann and C. Kittel : *Phys. Rev.* **106**, 893 (1957).

[7] W. Baltensperger and J. S. Helman : *Appl. Phys. Lett.* **57**, 2954 (1990).

[8] H. Sakakima and M. Satomi : *J. Magn. Magn. Mater.* **121**, 374 (1993).

[9] H. Watanabe, E. Hirota, A. Okada, K. Hamada, I. Ishida, H. Sakakima, and M. Satomi : *J. Phys. Soc. Jpn.* **63**, 762 (1994).

[10] H. Sakakima and M. Satomi : *Jpn. J. Appl. Phys.* **31**, L484 (1992).

[11] M. Jimbo, T. Kanda, S. Goto, S. Tsunashima and S. Uchiyama : *Jpn. J. Appl. Phys.* **31**, 1348 (1992).

[12] A. P. Malozemoff : *J. Appl. Phys.* **63**, 3874 (1988).

[13] W. Folkerts : *J. Magn. Magn. Mater.* **94**, 302 (1991).

[14] H. Sakakima, M. Satomi, K. Onaka, and S. Yamamoto : *J. Magn. Magn. Mater.* **145**, L268 (1995).

[15] Y. Saito, K. Inomata : *Jpn. J. Appl. Phys.* **30**, L1733 (1991).

[16] H. Kubota and T. Miyazaki : *J. Magn. Soc. Jpn.* **18**, 335 (1994).

[17] J. Inoue, A. Oguri and S. Maekawa : *J. Magn. Magn. Mater.* **104-107**, 1883 (1992).

[18] Y. Fujii, T. Ohnishi, T. Ishihara, Y. Yamada, K. Kawaguchi, N. Nakayama and T. Shinjo : *J. Phys. Soc. Jpn.* **55**, 251 (1986).

[19] L. Neel : *C. R. Acad. Sci.* **255**, 1676 (1962).

[20] Y. Kawawake, H. Sakakima, Y. Irie and M. Satomi : *Jpn. J. Appl. Phys.* **33**, L1598 (1994).

[21] P. Bruno and C. Chappert : *Phys. Rev. Lett.* **67**, 1602 (1991).

[22] M. T.Johnson, R. Coehoorn, J. J. de Vries, N. W. E. McGee, J. aan de Stegge, and P. J. H. Bloemen : *Phys. Rev. Lett.* **69**, 969 (1992).

[23] M. Satomi and H. Sakakima : *J. Magn. Magn. Mat.* **126**, 504 (1993).

[24] H. Sakakima, M. Satomi, Y. Irie, and Y. Kawawake : *Trans. Mat. Res. Soc. Jpn.* **15B**, 1083 (1994).

[25] K. Onaka, H. Sakakima, S. Yamamoto, S. Maekawa, M. Satomi and M. Okamoto: *National Tech. Rep.* **42**, 465 (1996).

4 Spin-Valve Devices

4.1 Spin Valves

Spin-alves are composed of a free and a pinned layer separated by a nonmagnetic layer. The spin-valve structure allows low-field GMR and the application of the spin-valves to magnetic read heads. MR properties of two types of spin-valves, exchange-biased spin valves having an antiferromagnetic pinning layer and pseudo-spin-valves having a hard magnetic layer, are discussed in this section.

4.1.1 Exchange-Biased Spin Valves

Exchange coupled metallic multilayers mentioned in Sect. 3.1 show large MR ratios, but the low-field sensitivity is not sufficiently high for the application of the multilayers to magnetic read heads. Spin valves proposed by Dieny *et al.* [1] realized high sensitivity GMR films for small magnetic fields. The single (or simple) spin-valve structure is composed of a free magnetic layer (FL), a nonmagnetic spacer layer (NM), a pinned magnetic layer (PL), and an antiferromagnetic pinning layer (AF) as shown in Fig. 4.1. The FL is usually composed of soft magnetic materials so that it can change the magnetic direction with an application of a small magnetic field. The NM is a separator to eliminate the magnetic coupling between the FL and the PL and to realize free magnetic rotation the in FL. The AF magnetically couples with the PL across the interface and causes the exchange bias effect on the PL. Figure 4.2 shows schematic curves of (a) magnetization and (b) magnetoresistance (MR) for spin-valve structured films. In the case where the pinning direction of the PL is the negative field direction, the configuration of the magnetization directions of the PL and the FL is parallel for H<0, which corresponds to the low resistance state. With the

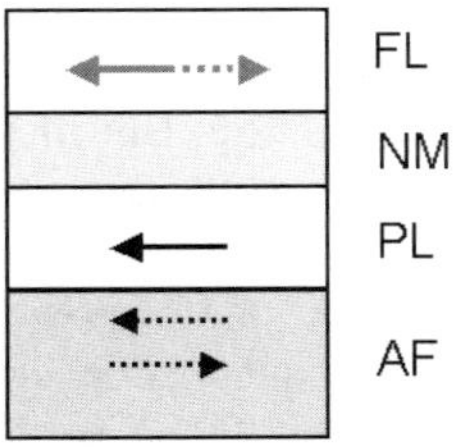

Fig. 4.1. Structure of the single (simple) spin-valve film

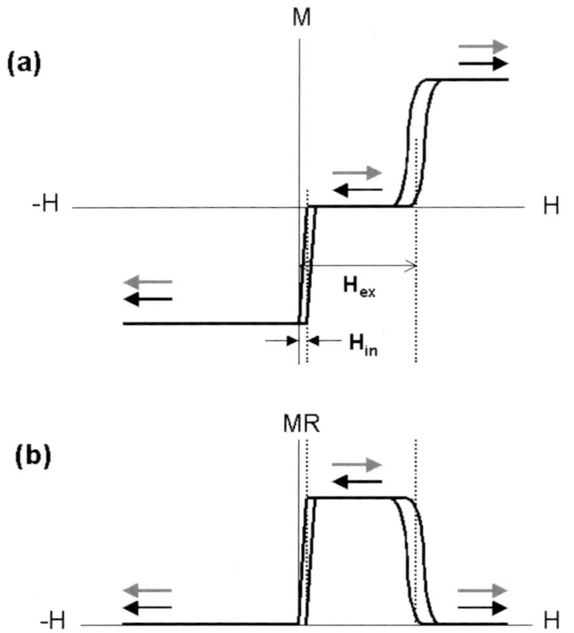

Fig. 4.2. **(a)** Magnetization curve and **(b)** MR curve for the single spin valve, where the magnetization directions of the FL and the PL are denoted by gray and black arrows, respectively

application of a small positive magnetic field, the magnetization direction of the FL switches from the negative to positive field direction, while that of the PL remains fixed, resulting in an antiparallel magnetization configuration that corresponds to the high resistance state. Thus, spin-valve films show a steep slope in the magnetoresistance curve (hereinafter, MR curve) in the vicinity of H=0 as shown in Fig. 4.2(b) and enable high sensitivity magnetic sensors and heads tobe made. With the application of a magnetic field larger than the exchange biased field, H_{ex}, the magnetization direction of the PL switches from the negative to positive field direction, resulting in a parallel magnetization configuration that corresponds to the low resistance state. Sometimes, there exists a small magnetic coupling between the FL and the PL, leading an offset field, H_{in}, for the magnetic switching of the FL as shown in the figure.

The prototype of the single spin-valve film is $Ni_{80}Fe_{20}$/Cu/$Ni_{80}Fe_{20}$/$Fe_{50}Mn_{50}$ [1]. The MR ratio is 2~5% at RT. This value is the same as, or about twice, that of conventional AMR (anisotropic magnetoresistance) films such as NiFe. But the thickness of the NiFe FL in the spin valve is thinner than 10 nm, while that of conventional AMR films is about 50 nm. Therefore, this spin-valve film offers higher sensitivity than conventional AMR films. Substitution of Co or CoFe for NiFe increases the MR ratio and enables higher output, but at the same time, it increases the coercivity of the FL and degrades the sensitivity. Therefore Co or CoFe is used for the PL and a thin Co or CoFe layer is inserted between the Cu spacer and the FL composed of NiFe to obtain higher MR ratios without degrading the sensitivity. FeMn shows poor corrosion resistance and low blocking temperature, T_B (~150 °C). These properties are improved by using PtMn [2] for the AF of the spin valves. A typical example of the single spin-valve film using PtMn is PtMn/CoFe/Cu/CoFe/NiFe. These PtMn-based spin valve films show better corrosion resistance, higher T_B (~380 °C) and large MR ratios (~8%). The characteristics of spin valves using various types of AF are summarized in Table 4.1, where the critical thickness is the minimum thickness for the AF films to show the exchange bias effect on the pinned layers and the blocking temperature is the temperature above which the AF films lose the pinning effect on the pinned layers.

NiMn and PtMn based spin valves need an annealing process at around 280 °C to obtain the ordered alloy of NiMn or PtMn to show the exchange bias effect, while FeMn and IrMn based spin valves do not need the annealing process. IrMn-based spin valves [9] show good corrosion resistance and large MR ratios (~9%). The critical thickness of IrMn is very thin compared with those of the AF used for spin valves. This is advantageous for the application of spin valves to thin film heads having narrow shield gaps. However, its blocking temperature, T_B, is not

Table 4.1. Characteristics of spin valves using various types of antiferromagnets

AF films	FeMn	IrMn	NiMn	PtMn	NiO	α-Fe_2O_3
Critical thickness(nm)	7~10	5~8	<25	10~15	<50	<50
Neel temp. (°C)	230	420	797	702	250	680
Blocking Temp. (°C)	150	250	370	380	200	250
Exchange bias field (Oe)	420	450	450~860	500	200	40~75
Corrosion Resistance	×	○	○	○	○	○
References	[3]	[4]	[5]	[6]	[7]	[8]

sufficiently high for practical application to GMR heads. Because the current density of the read head is as high as ~50 MA/cm^2 and the reading operation is done in a considerably high temperature environment, spin valve films are required to have high T_B, >300°C. Details of IrMn-based spin valves will be discussed in Sect. 4.2.1. The exchange bias fields and the T_B of NiO [10] and α-Fe_2O_3 [11] spin valves are not sufficiently high for practical application to GMR heads. But they have several excellent characteristics. They show very large MR ratios, 18~19% for the single spin valve structure and 23~28% [12-14] for the dual spin valve structure, which could not be obtained with spin valves using metallic AF films. Specular scattering of the conduction electrons at the interface between the oxide AF and PL is considered to contribute to the enhancement of the GMR. NiO and α-Fe_2O_3 could be used as the insulator between the spin valve and the shield layer in the read heads. This will be advantageous for application to GMR heads having narrow shield gap, if the critical thickness of NiO or α-Fe_2O_3 becomes less than 30 nm [15]. The details of α-Fe_2O_3 based spin valves will be discussed in Sect. 4.2.2. Generally speaking, these oxide AF films are used for "bottom spin valves" having the AF at the bottom, and metallic AF films are used for "top spin valves" having the AF on the top. But PtMn and IrMn films are used for both types of spin valves.

It is possible to increase the exchange bias field by using a synthetic antiferromagnetic (SAF) [16] pinned layer. In this SAF spin valve, the pinned layer is composed of antiferromagnetically exchange coupled magnetic layers, PL1 and PL2, across the nonmagnetic layer, NM*, as shown in Fig. 4.3. The FL is magnetically separated from the SAF pinned layer with NM as in the case of a simple spin valve and 2~3nm thick Cu is used for the NM. Co or CoFe is used for PL1 and PL2, and Ru or Ir is used for NM*. The sign of the interlayer exchange coupling energy between the PL1 and PL2 oscillates with the thickness of NM* as mentioned in Sect. 3.1.1. For example, strong antiferromagnetic coupling is realized at around 0.4 and 0.7 nm thick Ru [17] and at around 0.9~1 nm thick Cu [18]. It is possible to use 0.9 nm thick Cu as the NM*, but the thermal stability of the Co/Cu(0.9) interface is not so stable as that of Co/Ru(0.7). A typical example

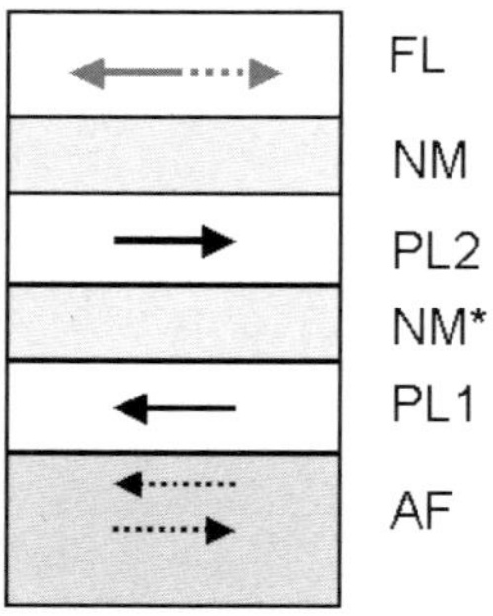

Fig. 4.3. Structure of the SAF spin valve

of SAF spin valve is PtMn/CoFe/Ru/CoFe/Cu/CoFe/NiFe which shows large exchange bias fields of 1~2 kOe. The SAF spin-valve structure is effective for improving the temperature dependence of the exchange bias field, though it could not increase the blocking temperature. In the simple spin valve, a stray field from the PL causes a bias effect on the FL, when the spin-valve film is patterned into a small element. The SAF spin-valve structure is desirable to reduce this bias effect as the stray field from the SAF pinned layer is very small so long as the difference in the thickness between the PL1 and PL2 is small, because the magnetization directions of the PL1 and PL2 are antiparallel. These are the reasons why SAF spin valves are suitable for practical application to GMR heads.

The SAF spin valves show a complicated MR curve as shown in Fig. 4.4. The magnetization direction of the AF layer in contact with the PL1 is assumed to be the positive field direction in this figure. Then, the magnetization direction of the PL1 and the AF at the interface tend to take a parallel magnetization configuration due to the exchange coupling, so long as the applied field is small. On the other hand, the PL2 and PL1 tend to take an antiparallel magnetization configuration due to the antiferromagnetic exchange coupling across the NM*. As a result, the magnetization direction of the PL2, which is antiparallel to that of the AF, is the negative field direction in this case, so long as the external magnetic field is small. With the application of a small *H*>0, the FL and the PL2 take a antiparallel magnetization configuration after the magnetization reversal of the FL Therefore,

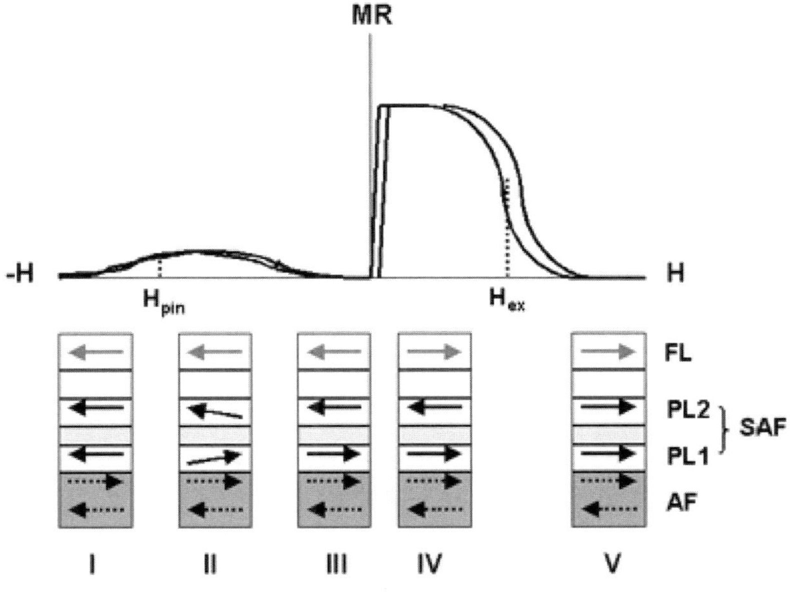

Fig. 4.4. MR curve of the SAF spin valve

the spin valve shows high resistance in this region IV. With the application of a larger H>0, all the magnetization directions of the magnetic layers become parallel after the magnetization reversal of PL2. Then, the spin valve shows a low resistance in region V. With the application of a small *H*<0, the FL and PL2 take a parallel magnetization configuration, while the PL2 and the PL1 keep the antiparallel magnetization configuration. If Cu is used for NM and Ru is used for NM*, the spin dependent scattering effect is dominant at the interface of FL/NM and of NM/PL2, but this is not dominant either at the interface of PL2/NM* nor at NM*/PL1, as the electron mean free path in Ru is very small compared with that in Cu. Then the spin valve shows a low resistance in this region III. With the application of a larger *H*<0, the magnetization directions of the PL1, the PL2 and the FL the magnetic layers become parallel after the magnetization reversal of PL1 and the spin valve shows a low resistance in the region I. Midway between region III and region I, a spin flop occurs in the PL1 and the PL2 and a small increase of the resistance is observed in region II. The magnetization reversal of PL2 with $H>0$ is done against the antiferromagnetic exchange coupling with PL1 and the threshold field is defined as H_{ex}, while the magnetization reversal of PL2 with $H<0$ is done against the exchange coupling with AF as well as the antiferromagnetic exchange coupling with PL1 and the threshold field is defined as H_{pin}.

Spin-dependent scattering of the conduction electrons occurs at the interfaces between the nonmagnetic layers and the magnetic layers. Therefore, it is possible to increase the MR ratios by increasing the number of interfaces so long as the total film thickness is smaller than the mean free path of the conduction electrons. The dual (or symmetric) spin valve structure [19] enables one to obtain larger MR ratios than the single spin-valve structure. Dual spin valves are composed of two PL (PL-t and PL-b), two NM (NM-t and NM-b), and a FL as shown in Fig. 4.5. The FL is magnetically separated from the PL-t and PL-b with the NM-t and NM-b inserted between the FL and PL-t and PL-b. Spin-dependent scattering of the electrons occurs at the interfaces of PL-t/NM-t, NM-t/FL, FL/NM-b, and

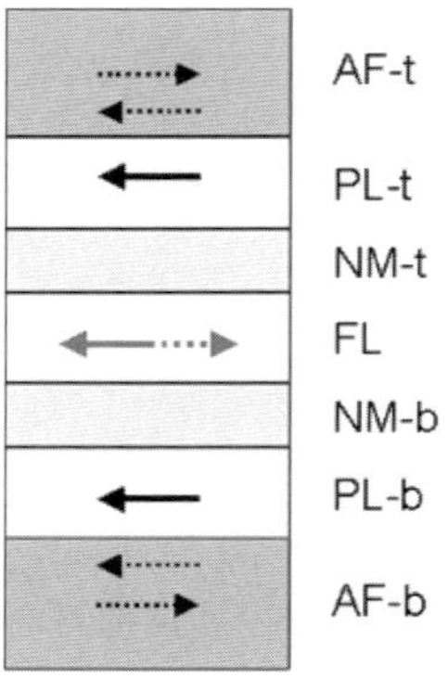

Fig. 4.5. Structure of the dual (or symmetric) spin valve

NM-b/PL-b in dual spin valves, while it occurs only at the interfaces of FL/NM and NM/PL in the single spin valves shown in Fig. 4.1. A typical example of a dual spin valve is UL/PtMn/CoFe/Cu/CoFe/NiFe/CoFe/Cu/CoFe/PtMn, where UL is a seed layer which helps the fcc structure growth of the PtMn layer and Ta is usually used for the UL for this purpose. In this example, the FL is composed of CoFe/NiFe/CoFe, because the CoFe/Cu interface is effective for obtaining large MR ratios and the NiFe is effective for improving the soft magnetic property of the FL. SAF dual spin valves are also studied for PtMn or IrMn based spin valves [20,21]. A typical example of the SAF dual spin valve is UL/AF/CoFe/Ru/CoFe/Cu/CoFe/NiFe/CoFe/Cu/CoFe/Ru/CoFe/AF (AF=PtMn or IrMn). These dual spin valves enable one to obtain larger MR ratios (~12%) than those of single spin valves. However, the total film thickness of the dual spin valve is larger than that of the single spin valve. Therefore, dual spin-valve structure is disadvantageous for the application to GMR heads having an extremely narrow shield gap.

4.1.2 Pseudo-Spin-Valves

Exchange-biased spin valves use antiferromagnets. There is another type of spin valves called a "pseudo-spin-valves" [22] that do not use antiferromagnets.

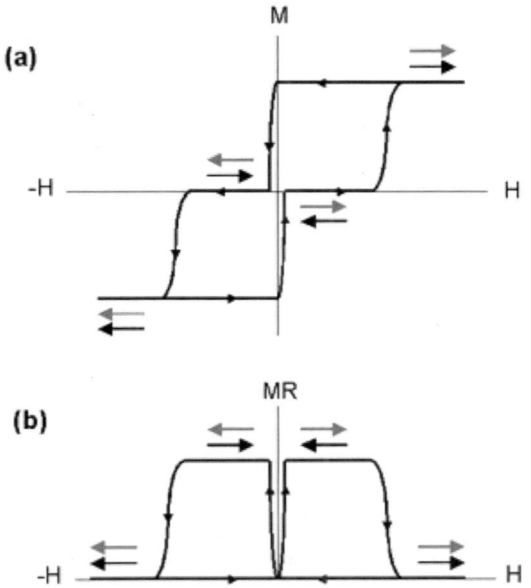

Fig. 4.6. **(a)** Magnetization curve and **(b)** MR curve of the pseudo-spin-valve, where the magnetization directions of the SM and the HM are denoted with gray and black arrows, respectively

The pseudo-spin-valve is composed of a hard magnetic layer, HM, and a soft magnetic layer, SM, magnetically separated with a nonmagnetic layer, NM. The HM has a larger coercivity than the SM. The schematic magnetization curve and the MR curve of the pseudo-spin-valve are shown in Fig. 4.6 (a) and (b). With the application of a small magnetic field that causes magnetization reversal of the SM, but not of the HM, the magnetization directions of the SM and the HM take an antiparallel configuration. Then, the pseudo-spin-valve shows high resistance. With the application of a large magnetic field that causes magnetization reversal of the HM, the magnetization directions of the SM and HM take a parallel configuration and the pseudo-spin-valve shows low resistance. Pseudo-spin-valves shows a symmetric MR curve with the magnetic field direction, while the exchange-biased spin valves show a nonsymmetric one as shown in Fig. 4.2(b), as the HM in the pseudo-spin-valve has uniaxial anisotropy, while the PL/AF in the exchange-biased spin valve has unidirectional anisotropy. The exchange-biased spin valve shows a reversible MR curve with the magnetic field. But the pseudo-spin-valve shows an irreversible one. This is a disadvantage of pseudo-spin-valves for application to MR heads. However, pseudo-spin-valves are studied for application to MRAM (magnetic random access memory) devices [23,24] which will be discussed later.

A prototype pseudo-spin-valve film proposed by Shinjo *et al.* is $[\mathrm{NiFe}(3)/\mathrm{Cu}(5)/\mathrm{Co}(3)/\mathrm{Cu}(5)]^{15}$ multilayered film [25]. NiFe is used for the SM and Co is used for the HM. This multilayered film showed a MR ratio of ~8%. Epitaxial film growth of $[\mathrm{NiFe/Cu/Co\ /Cu}]^{N}$ multilayered films on Si(100) with a Cu buffer layer using MBE is also studied [26]. Figure 4.7 shows an MR curve of epitaxially grown [NiFe(10)/Cu(5)/Co(5)/Cu(5)/NiFe(10)] film measured along [011] in the

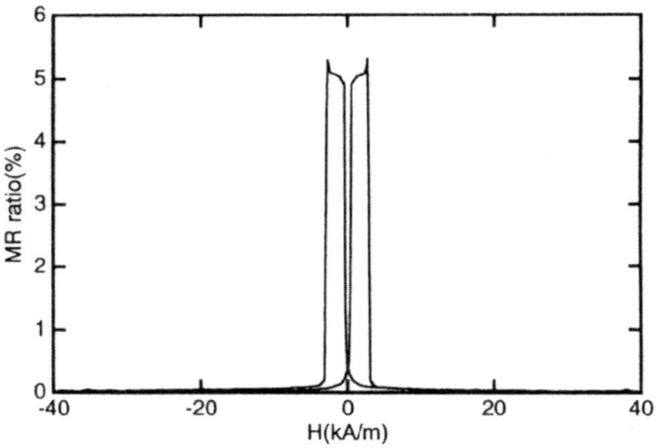

Fig. 4.7. MR curve of the [NiFe/Cu/Co/Cu/NiFe] pseudo-spin-valve epitaxially grown on Si(100)

(100) plane. The magnetocrystalline anisotropy energy, E_a, for the films having cubic symmetry is given as

$$E_a = K_1/8(1-\cos 4\theta) + \cdots \tag{4.1}$$

when the magnetization rotates in the (100) plane, where θ is the angle between the magnetization direction and the [010] axis. The sign of K_1 is negative for fcc NiFe and Co. Therefore, the NIFe and Co layers have the easy axis along the [011] direction. This is the reason why the MR curve measured along the [011] direction shows a steep slope where the magnetization reversal of the NiFe and Co layers take place. These characteristics are suitable for the application of pseudo-spin-valves to MRAM devices [27], as the magnetization reversal of the HM (=Co) layers is used for the data writing and that of SM(=NiFe) layers is used for data reading. The switching field of SM is lower than that of HM. So the data reading is done without the magnetization reversal of the HM layers. Therefore, NDR (nondestructive read-out) is possible for this pseudo-spin-valve memory [22]. A large MR ratio, ~10%, with a small switching field, ~10 Oe, of SM

(=NiFe) is obtained for epitaxially grown [Co(3)/Cu(2.4)/Ag(0.2)/Cu(2.4)/NiFe(3)/Cu(2.4)/Ag(0.2)/Cu(2.4)]^{10} mulltilayered film [28]. The ultra-thin Ag layers are inserted in the Cu layer so as to reduce the magnetic coupling between the SM=NiFe and the HM=Co layers.

Various types of pseudo-spin-valves using hard magnetic materials as the HM layers or using soft magnetic amorphous alloys as the SM layers are also studied. Examples of the former type are CoPt(3)/Cu(2.1)/NiFeCo(9.8) [29] which shows high corrosion resistance and $CoFe_2O_4$(50)/CoFe(2)/Cu(2.1)/CoFe(5) [30] that

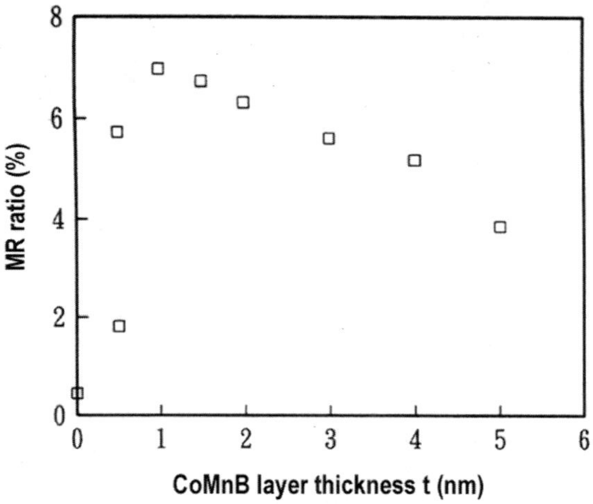

Fig. 4.8. Dependence of the MR ratio on the CoMnB layer thickness in CoMnB(t)/Co(1.2)/Cu(2.4)/Co(2)

shows large MR ratios owing to the specular scattering of the conduction electrons at the $CoFe_2O_4$/CoFe interface. Details will be discussed in Sect. 4.2.2. Examples of the latter type are CoFeB(6)/Cu(2.1)/Co(1.5) [31] and CoMnB(1.5)/Co(1.2)/Cu(2.1)/Co(2) [32]. These spin valves enable one to reduce the total film thickness of the spin valves less than 10 nm keeping considerably large MR ratios of 7~8%. These Co-based amorphous films show a soft magnetic property even when the film thickness is smaller than 2 nm, which is not easy to achieve with NiFe. The electron mean free path in the amorphous layer is much smaller than in NiFe or in Co. So increase of the amorphous layer thickness increases the resistance of the spin valves, resulting in a decrease of the MR ratios. A phenomenological expression for the MR ratio [33] is given as,

$$\Delta R/R=(\Delta R/R)_0[1\text{-}\exp(-t_F/l_F)/(1+t_F/t_0)] \tag{4.2}$$

where t_F and t_0 are the thickness of the ferromagnetic layer and a fitting parameter, respectively, and l_F is expected to be about twice that of the mean free path of the majority electrons. Therefore, the maximum MR ratio is obtained with a very thin (~1 nm) amorphous layer, as shown in Fig. 4.8. As a result, these spin valves have very thin SM and small demagnetization fields, which is advantageous when they are patterned into small elements as the demagnetization field increases with the reduction of the size, if the thickness is not reduced.

4.2 Enhancement of GMR

GMR is a phenomenon caused by spin-dependent scattering of conduction electrons in the magnetic layers and at the interfaces between the magnetic and the nonmagnetic layers. Therefore, the MR ratios depend upon the relation between the electron mean free path, MFP, and the layer thickness of the magnetic and the nonmagnetic layers of the GMR films. It sometimes happens that the thickness of the GMR film is smaller than the MFP. In this case, enhancement of GMR is possible to increase the effective MFP for the parallel spin configuration by the back-layer effect or by the specular scattering effect.

4.2.1 Back-Layer Effect

The back-layer effect was first described by Gurney *et al.* [34]. They studied the conductance of the exchange-biased spin valve having a back layer in contact with the FL which is assumed to be thinner than the mean free path for majority spin electrons. In this case, enhancement of GMR is possible by increasing the FL thickness, or by adding a "back layer", BL, on the FL as shown in Fig. 4.9. The spin valve has a structure of UL(Ta/Cu)/AF(FeMn)/PL(NiFe)/NM(Cu)/FL(NiFe)/BL/CL(Ta), where UL is a under layer and CL is a capping layer. The back layer is expected to increase the effective mean free path of the majority spin electrons for the parallel magetization alignment. Various metallic layers such as Cu, AuCu,

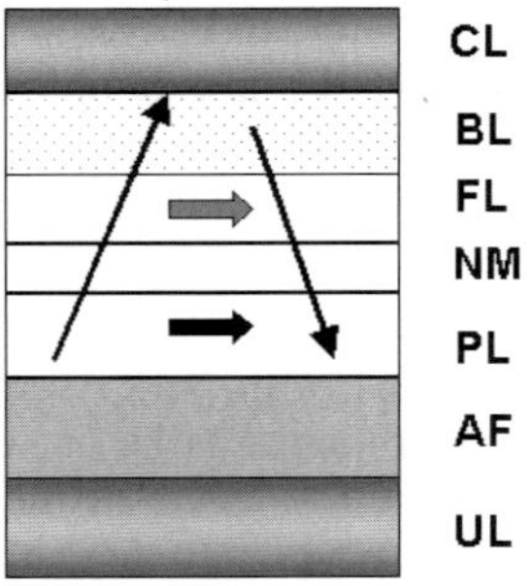

Fig. 4.9. Schematic trajectories of majority spin electrons in the spin valve with a back layer

Co, NiFe, and Fe were studied as the back layer. The addition of the back layer increases the conductance for parallel magnetization alignment, but not for the antiparallel alignment. Because the electrons in the PL could not penetrate into the back layer due to the spin filtering effect of the FL having the opposite spin state against the PL for the antiparallel magnetization alignment, the difference between the conductance for parallel magnetization alignment, G_p, and that for antiparallel alignment, G_{ap}, is expected to increase with the addition of the back layer. In reality, $\Delta G(=G_p-G_{ap})$ was found to increase, to a good approximation, following the relation [34]

$$\Delta G=\Delta G_0+\Delta G_B[1-\exp(-t_B/l_B)] \tag{4.3}$$

where t_B is the thickness of the back layer and the parameter l_B is estimated to be about equal to the mean free path of the majority spin electrons in the back layer. For example, l_B for Cu, Co, NiFe, Fe was 19.2, 5.5, 4.6, and 1.5 nm, respectively. So the addition of a metallic layer, such as Cu, having a long electron mean free path, on the FL was found to be effective in increasing ΔG owing to the spin filter effect of the FL, though the addition of a metallic buffer layer under the AF simply reduced the MR ratios due to the shunting effect.

The back layer effect implies the possibility of the enhancement of GMR with the addition of a high conduction layer, HCL, in contact with the FL, as the MR ratio is given as $\Delta G/G_{ap}$. This was achieved by the Tohshiba group using the "spin-filter spin valve" [35] composed of UL(Ta)/HCL/FL(CoFe)/NM(Cu)/PL (=SAF)/AF (IrMn)/CL(Ta), where SAF is composed of CoFe/Ru/CoFe) as shown in Fig. 4.10. The sensitivity of the spin-valve elements depends upon the product of the magnetization and the thickness of the FL. So it is desirable to reduce the FL thickness. But, at the same time, it increases the sensitivity of the FL against the bias effect of the stray field from the PL. The reduction of the FL thickness causes a decrease of the MR ratio and a deterioration of the soft magnetic property of a FL in conventional spin valves. These problems are almost solved by using spin-filter spin-valve structure. The stray field bias is easily reduced by the adoption of the SAF structure to the PL, though it causes a decrease of the MR

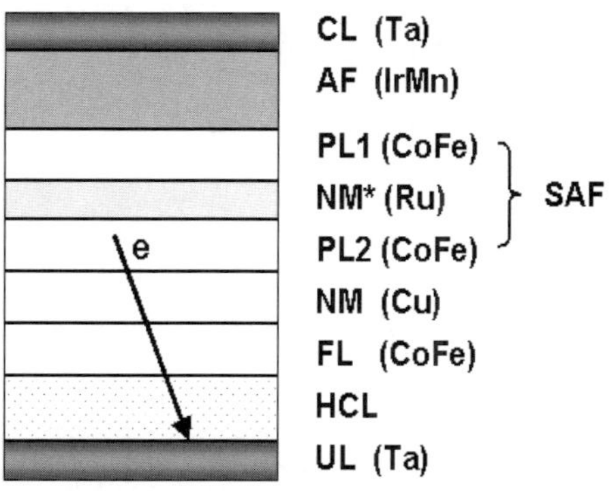

Fig. 4.10. Structure of the spin-filter spin valve

ratio to some extent, because the electron mean free path in Ru is much smaller than in CoFe, which results in the reduction of the G_p of the spin valve. The addition of the HCL increases the G_p and the MR ratio of the spin valve owing to the back-layer effect. The MR ratio of the spin-filter spin valve can be simulated by using Barnas's model [36]. Figure 4.11 shows the numerical results of the dependence of the MR ratios on the FL thickness, t_{FL}, for SFSV (spin-filter spin valve) having a structure of PL(5)/NM(2)/FL(t)/HCL(3) and for CSV (conventional spin valve) having a structure of PL(5)/NM(2)/FL(t), which qualitatively explains the experimental results for the SFSV [35]. It turns out that the SFSV enables one to obtain a larger MR ratio than the CSV when the FL is thinner

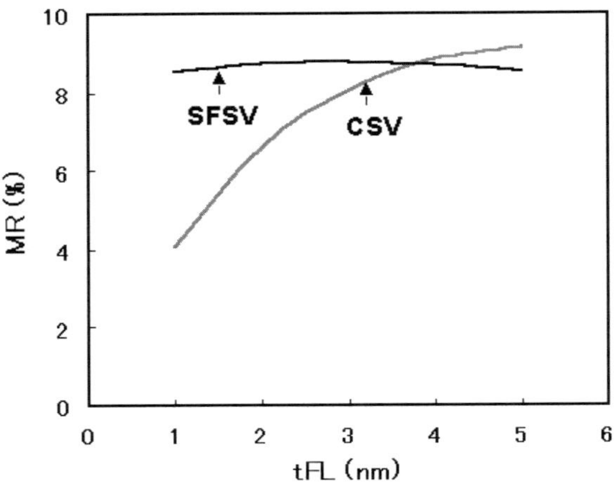

Fig. 4.11. Dependence of the MR ratios on the FL thickness for SFSV and CSV

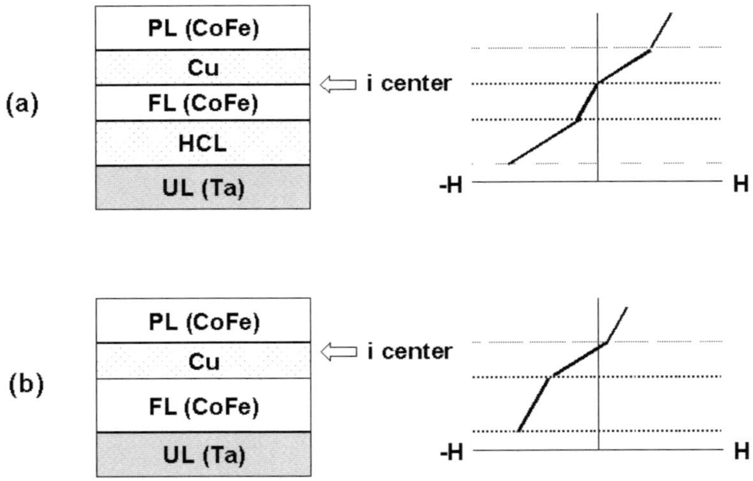

Fig. 4.12. Sense current field distribution in **(a)** SFSV and **(b)** CSV

than 3 nm. The addition of the HCL is also effective to adjust the bias field on the FL generated by the sense current. The current center is in the NM in the CSV and shifts towards the FL with the addition of HCL as shown in Fig. 4.12. Therefore the bias effect of the sense current on the FL is smaller in the SFSV than in the CSV. The soft magnetic property of the thin FL(=CoFe) is improved by the film growth on the UL(=Ta)/HCL(=Cu). The Ta/Cu seed layer helps the fcc (111) texture growth of the CoFe and contribute the reduction of the coercivity, H_C, and the magnetostriction, λ, of CoFe. For example, H_C of CoFe was >20 Oe on Ta, 10~15 Oe on Cu, and 3~4 Oe on Ta/Cu, respectively. The characteristics of the SFSV are summarized in Table. 4.2, where H_K is the anisotropy field of the FL and H_{in} is the exchange coupling field between the PL and the FL. The SFSV shows excellent characteristics for the practical application to GMR heads. PtMn-based SFSV is also studied to improve the T_B [37].

Table 4.2. Characteristics of SFSV

Properties	Values
MR (%)	9 ~ 10
ΔRs (Ω/□)	1.5 ~ 2
Rs (Ω/□)	16 ~ 20
Hc (Oe)	~ 3
Hk (Oe)	~ 8
λ (ppm)	≤ 0.5
Hin (Oe)	≤ 10
T_B (°C)	~ 270

4.2.2 Specular Scattering Effect

The intrinsic mean free path, MFP, of the conduction electron in the spin valves is reduced because of the geometrical boundary, where the diffusive scattering of the

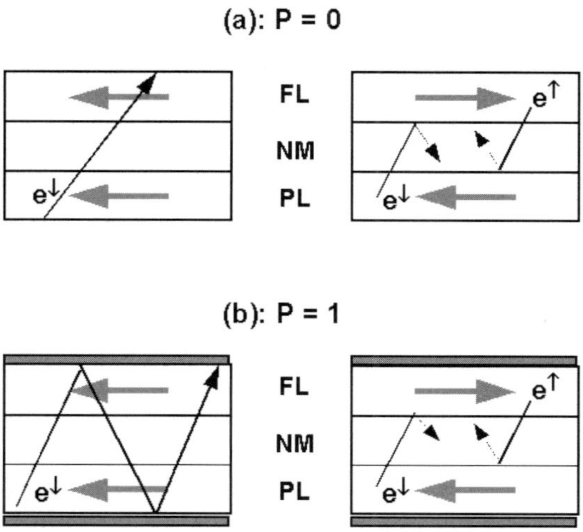

Fig. 4.13. Schematic illustration of the electron trajectories in spin valves for **(a)** diffusive scattering and **(b)** specular scattering

conduction electron is dominant as shown in Fig. 4.13(a). On the contrary, the exchange coupled GMR having the film thickness larger than the MFP, such as $[Co/Cu]^N$ multilayers, show large MR ratios of ~100% as mentioned in Sect. 3.1. If both the top and bottom surfaces of the spin valves were perfectly specularly reflective (*ie.* specularity factor, P=1), the effective MFP could be as long as the intrinsic one in the case of the parallel magnetization alignment as shown in Fig. 4.13(b). Then the G_p would be much larger than G_{ap}, resulting in the enhancement of the GMR. The specular scattering effect at the interface between the PL and the insulative AF, such as Co/NiO, was first described by Swagten *et al.* [38] and enhanced GMR (>21%), by specular scattering was reported in NiO/Co/Cu/Co/Cu/Co/NiO dual spin valves by Egelhoff *et al.* [12]. A larger GMR of ~28% was reported in α-Fe_2O_3/Co/Cu/Co/Cu/Co/α-Fe_2O_3 dual spin valves by Sakakima *et al.* [13]. Specular scattering at the interface between the vacuum and the capping layer, CL, such as Au and Ag, was also found in NiO/Co/Cu/Co/Au bottom spin valves [39] and NiFe/Cu/Co/Cu/Ag pseudo-spin-valves [40]. Recently, the enhancement of the GMR by specular reflective thin oxide layers, NOL, inserted in the PL or/and FL was reported in IrMn and PtMn based spin valves by the Tohshiba and Matsushita group [41,42].

4.2.2.1 Specular Scattering by Capping Layers

First, let us consider the enhancement of GMR by specularly reflective CL in a NiFe/Cu/Co pseudo-spin-valve prepared on MgO(001) with Pt/Cu under layers as shown in Fig. 4.14. With the addition of a Cu capping layer on the Co layer, the MR ratio decreased from 4.5% to 3.9%, due to the shunting effect. However, further addition of an Ag capping layer on the Cu layer enhanced the MR ratios from 3.9% to 5.4%, though it also increased the G_p due to the shunting effect. This result may be due to specular scattering at the Ag/vacuum interface. Figure 4.15 shows the dependence of the MR ratios and ΔG on the Ag layer thickness, t_{Ag}. Discrepancies were found in the MR ratio and ΔG between t_{Ag}=0 and t_{Ag}=1 nm and the values showed abrupt increases with the addition of 1nm thick Ag.These jumps in the MR ratio and ΔG imply a discontinuous increase in the specularity at the top surface of the spin valve caused by changing the CL from Cu to Cu/Ag.

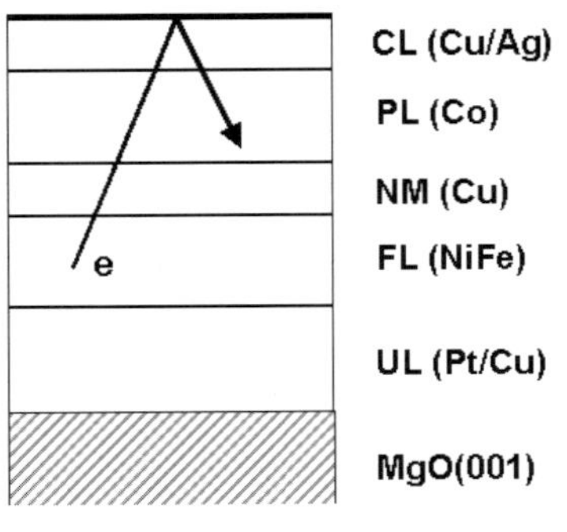

Fig. 4.14. Structure of the pseudo-spin-valve having a specular reflective CL

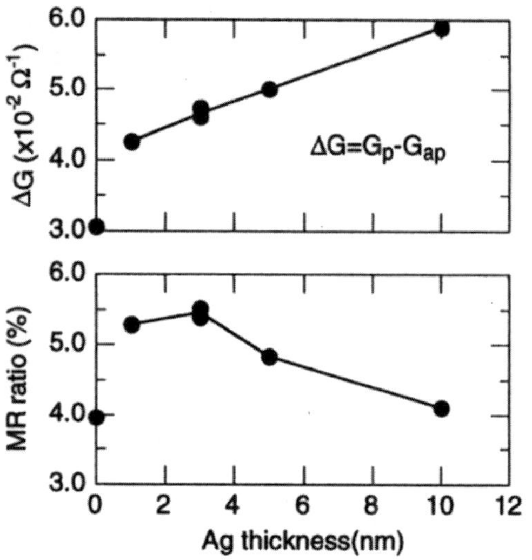

Fig. 4.15. Dependence of the MR ratios and ΔG on the Ag layer thickness

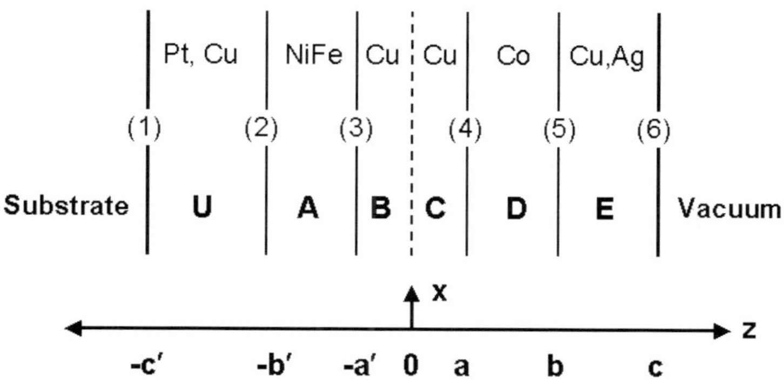

Fig. 4.16. Geometry of the spin-valve and the coordinate system used for the simulation

It is possible to simulate the MR ratios of the spin valves using a theoretical model given by Barnas *et al.* [36]. In an electric field, E, the shift of the spin-dependent distribution function of the electrons with velocity v on the Fermi surface from thermal equilibrium, $G(z,v)$, is expressed as follows:

$$G_{\pm}^{\uparrow\downarrow}(z,v)=(eE\tau^{\uparrow\downarrow}/m)(\partial f_0/\partial v_x)[1+F_{\pm}^{\uparrow\downarrow}(v)/\exp\{\mp z/(\tau^{\uparrow\downarrow}|v_z|)\}], \qquad (4.4)$$

where f_0 denotes the Fermi distribution function and ± refers to electrons moving towards ± directions of the z-axis and the relaxation time, τ, is given using the electron MFP, λ, as $\tau^{\uparrow\downarrow}=\lambda^{\uparrow\downarrow}/v$. The function $F(v)$ is an arbitrary function which is determined by the boundary conditions for scattering characteristic of electrons at the interfaces and surfaces of the film. The MR ratios of the [NiFe/Cu/Co] spin valves having Cu/Ag capping layers and Pt/Cu under layers are calculated with the model shown in Fig. 4.16, where the z-axis is along the normal to the layers and the x-axis is in the plane of the layers and is parallel to both the applied electric and magnetic fields. Then the boundary conditions are given as:

at the bottom surface (1)

$$G_{U+}^{\uparrow\downarrow} = P_U G_{U}^{-\uparrow\downarrow} \tag{4.5}$$

at interface (2)

$$\left.\begin{aligned} G_{U}^{-\uparrow\downarrow} &= T^{\uparrow\downarrow} G_{A}^{-\uparrow\downarrow} + R^{\uparrow\downarrow} G_{U+}^{\uparrow\downarrow} \\ G_{A+}^{\uparrow\downarrow} &= T^{\uparrow\downarrow} G_{U+}^{\uparrow\downarrow} + R^{\uparrow\downarrow} G_{A}^{-\uparrow\downarrow} \end{aligned}\right] \tag{4.6}$$

at interface (3)

$$\left.\begin{aligned} G_{A}^{-\uparrow\downarrow} &= T^{\uparrow\downarrow} G_{B}^{-\uparrow\downarrow} + R^{\uparrow\downarrow} G_{A+}^{\uparrow\downarrow} \\ G_{B+}^{\uparrow\downarrow} &= T^{\uparrow\downarrow} G_{A+}^{\uparrow\downarrow} + R^{\uparrow\downarrow} G_{B}^{-\uparrow\downarrow} \end{aligned}\right] \tag{4.7}$$

at interface (4)

$$\left.\begin{aligned} G_{C}^{-\uparrow\downarrow} &= T^{\uparrow\downarrow} G_{D}^{-\uparrow\downarrow} + R^{\uparrow\downarrow} G_{C+}^{\uparrow\downarrow} \\ G_{D+}^{\uparrow\downarrow} &= T^{\uparrow\downarrow} G_{C+}^{\uparrow\downarrow} + R^{\uparrow\downarrow} G_{D}^{-\uparrow\downarrow} \end{aligned}\right] \tag{4.8}$$

at interface (5)

$$\left.\begin{aligned} G_{D}^{-\uparrow\downarrow} &= T^{\uparrow\downarrow} G_{E}^{-\uparrow\downarrow} + R^{\uparrow\downarrow} G_{D}^{+\uparrow\downarrow} \\ G_{E+}^{\uparrow\downarrow} &= T^{\uparrow\downarrow} G_{D+}^{\uparrow\downarrow} + R^{\uparrow\downarrow} G_{E}^{-\uparrow\downarrow} \end{aligned}\right] \tag{4.9}$$

at the top surface (6)

$$G_{E}^{-\uparrow\downarrow} = P_E G_{E+}^{\uparrow\downarrow} \tag{4.10}$$

and at z=0

$$\left.\begin{aligned} G_{C+}^{\uparrow\downarrow} &= \cos^2(\theta/2) G_{B+}^{\uparrow\downarrow} + \sin^2(\theta/2)\, G_{B+}^{\uparrow\downarrow} \\ G_{B}^{-\uparrow\downarrow} &= \cos^2(\theta/2) G_{C}^{-\uparrow\downarrow} + \sin^2(\theta/2)\, G_{C}^{-\uparrow\downarrow} \end{aligned}\right] \tag{4.11}$$

The fictitious boundary conditions of (4.11) correspond to the change in magnetization axis between the magnetic A and D layers, where θ is the angle between the magnetizations in the A and D layers. Therefore, θ=0 and π correspond to ferromagnetic and antiferromagnetic configurations, respectively. R and T are the reflection and transmission coefficients of the electrons at the interface; P_U and P_E are specularity factors at the surfaces of the U and E layers.

Equations (4.5~4.11) are transformed into the following equations:

at the bottom surface (1)

$$-C'_U F_{U+}^{\uparrow\downarrow} + (P_U/C'_U) F_{U}^{-\uparrow\downarrow} = 1 - P_U \tag{4.5'}$$

at interface (2)

$$\left.\begin{aligned} &R^{\uparrow\downarrow}B'_{U}F_{U+}{}^{\uparrow\downarrow} - (1/B'_{U})F_{U-}{}^{\uparrow\downarrow} + (T^{\uparrow\downarrow}Y_{AU}{}^{\uparrow\downarrow}/B'_{A}{}^{\uparrow\downarrow})F_{A-}{}^{\uparrow\downarrow} \\ &\qquad = 1 - R^{\uparrow\downarrow} - T^{\uparrow\downarrow}Y_{AU}{}^{\uparrow\downarrow} \\ &(T^{\uparrow\downarrow}Y_{UA}{}^{\uparrow\downarrow}B'_{U})F_{U+}{}^{\uparrow\downarrow} - B'_{A}{}^{\uparrow\downarrow}F_{A+}{}^{\uparrow\downarrow} + (R^{\uparrow\downarrow}/B'_{A}{}^{\uparrow\downarrow})F_{A-}{}^{\uparrow\downarrow} \\ &\qquad = 1 - R^{\uparrow\downarrow} - T^{\uparrow\downarrow}Y_{UA}{}^{\uparrow\downarrow} \end{aligned}\right\} \tag{4.6'}$$

at interface (3)

$$\left.\begin{aligned} &R^{\uparrow\downarrow}A_{A}{}^{\uparrow\downarrow}F_{A+}{}^{\uparrow\downarrow} - (1/A_{A}{}^{\uparrow\downarrow})F_{A-}{}^{\uparrow\downarrow} + (T^{\uparrow\downarrow}Y_{BA}{}^{\uparrow\downarrow}/A_{B})F_{B-}{}^{\uparrow\downarrow} \\ &\qquad = 1 - R^{\uparrow\downarrow} - T^{\uparrow\downarrow}Y_{BA}{}^{\uparrow\downarrow} \\ &(T^{\uparrow\downarrow}Y_{AB}{}^{\uparrow\downarrow}A_{A}{}^{\uparrow\downarrow})F_{A+}{}^{\uparrow\downarrow} - A_{B}F_{B+}{}^{\uparrow\downarrow} + (R^{\uparrow\downarrow}/A_{B})F_{B-}{}^{\uparrow\downarrow} \\ &\qquad = 1 - R^{\uparrow\downarrow} - T^{\uparrow\downarrow}Y_{AB}{}^{\uparrow\downarrow} \end{aligned}\right\} \tag{4.7'}$$

at interface (4)

$$\left.\begin{aligned} &(R^{\uparrow\downarrow}/A_{C})F_{C+}{}^{\uparrow\downarrow} - A_{C}F_{C-}{}^{\uparrow\downarrow} + (T^{\uparrow\downarrow}Y_{DC}{}^{\uparrow\downarrow}A_{D}{}^{\uparrow\downarrow})F_{D-}{}^{\uparrow\downarrow} \\ &\qquad = 1 - R^{\uparrow\downarrow} - T^{\uparrow\downarrow}Y_{DC}{}^{\uparrow\downarrow} \\ &(T^{\uparrow\downarrow}Y_{CD}{}^{\uparrow\downarrow}/A_{C})F_{C+}{}^{\uparrow\downarrow} - (1/A_{D}{}^{\uparrow\downarrow})F_{D+}{}^{\uparrow\downarrow} + (R^{\uparrow\downarrow}A_{D}{}^{\uparrow\downarrow})F_{D-}{}^{\uparrow\downarrow} \\ &\qquad = 1 - R^{\uparrow\downarrow} - T^{\uparrow\downarrow}Y_{CD}{}^{\uparrow\downarrow} \end{aligned}\right\} \tag{4.8'}$$

at interface (5)

$$\left.\begin{aligned} &(R^{\uparrow\downarrow}/B_{D}{}^{\uparrow\downarrow})F_{D+}{}^{\uparrow\downarrow} - B_{D}{}^{\uparrow\downarrow}F_{D-}{}^{\uparrow\downarrow} + (T^{\uparrow\downarrow}Y_{ED}{}^{\uparrow\downarrow}B_{E})F_{E-}{}^{\uparrow\downarrow} \\ &\qquad = 1 - R^{\uparrow\downarrow} - T^{\uparrow\downarrow}Y_{ED}{}^{\uparrow\downarrow} \\ &(T^{\uparrow\downarrow}Y_{DE}{}^{\uparrow\downarrow}/B_{D}{}^{\uparrow\downarrow})F_{D+}{}^{\uparrow\downarrow} - (1/B_{E})F_{E+}{}^{\uparrow\downarrow} + (R^{\uparrow\downarrow}B_{E})F_{E-}{}^{\uparrow\downarrow} \\ &\qquad = 1 - R^{\uparrow\downarrow} - T^{\uparrow\downarrow}Y_{DE}{}^{\uparrow\downarrow} \end{aligned}\right\} \tag{4.9'}$$

at the top surface (6)

$$-C_{E}F_{E-}{}^{\uparrow\downarrow} + (P_{E}/C_{E})F_{E+}{}^{\uparrow\downarrow} = 1 - P_{E} \tag{4.10'}$$

and at z=0

$$\left.\begin{aligned} &\cos^2(\theta/2)F_{B+}{}^{\uparrow\downarrow} + \sin^2(\theta/2)F_{B+}{}^{\uparrow\downarrow} - F^{\uparrow\downarrow}{}_{C+} = 0 \\ &\cos^2(\theta/2)F_{C-}{}^{\uparrow\downarrow} + \sin^2(\theta/2)F_{C-}{}^{\uparrow\downarrow} - F^{\uparrow\downarrow}{}_{B-} = 0 \end{aligned}\right\} \tag{4.11'}$$

where

$$C'_{U} = \exp(c'/\lambda_{U}\cos\bullet),\ B'_{A}{}^{\uparrow\downarrow} = \exp(b'/\lambda_{A}{}^{\uparrow\downarrow}\cos\beta), \text{ etc.,} \tag{4.12}$$

$$Y_{AU}{}^{\uparrow\downarrow} = (\lambda/mv)_{A}/(\lambda/mv)_{U} = 1/Y_{UA}{}^{\uparrow\downarrow}, \text{ etc.,} \tag{4.13}$$

and β is the angle between electron velocity and the z-axis. These F functions are determined as the functions of θ and β from these conditions. The electric current, I, is calculated using these F-functions as follows:

$$I = \Sigma I_U = \Sigma \int dv\{-ev_x(g^{\uparrow}+g^{\downarrow})\} = \sigma Et, \tag{4.14}$$

where σ and t are the conductivity and the total thickness of the layered film, respectively. The conductance $G(\theta) = \sigma t$ is calculated using a similar procedure to that given by Barnas *et al.*[36]. Then the MR ratio is obtained as $\Delta G/G_{ap}$ ($\Delta G = G_p - G_{ap}$).

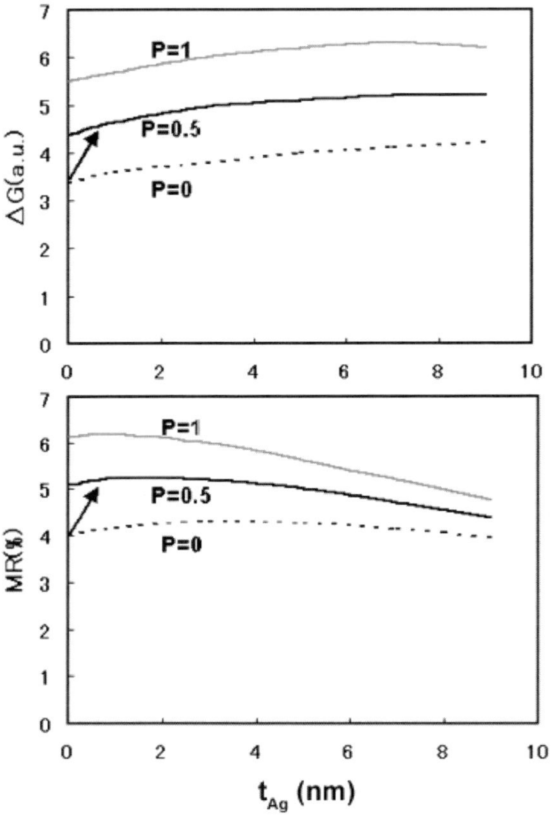

Fig. 4.17. Calculated results for the dependence of ΔG and the MR ratio on t_{Ag} and P of the capping layer

The dependence of the MR ratio and ΔG on the Ag capping layer thickness, t_{Ag}, and the specularity factor of the capping layer, $P(=P_E)$, in Pt(10 nm)/Cu(5 nm)/NiFe(5 nm)/Cu(2.8 nm)/Co(5 nm)/Cu(1.2 nm)/Ag(t_{Ag} nm) are calculated using this model. The calculation was done with the following assumption: The nonmagnetic layers have the same MFP, λ_2(=60 nm), and ferromagnetic layers have the same spin-dependent MFP, $\lambda_1=(\lambda_1^{\uparrow}+\lambda_1^{\downarrow})/2$(=15 nm) with spin asymmetry of $N_B=\lambda_1^{\downarrow}/\lambda_1^{\uparrow}$(=2). $T^{\downarrow}$, $T^{\uparrow}$ and R are 0.8, 0.2, and 0, respectively at the interfaces, and P_U is 0 at the bottom surface. Figure 4.17 shows the dependence of ΔG and the MR ratio on t_{Ag} and P of the capping layer, N_3, in N_1(7 nm)/M_1(5 nm)/N_2(2.8 nm)/M_2(5 nm)/N_2(1.2 nm)/N_3(t_{Ag} nm), where M and N denote the magnetic and nonmagnetic layers, respectively. N_1, M_1, N_2, M_2, and N_3 correspond to Pt/Cu, NiFe, Cu, Co and Ag layers of the spin valve, respectively.

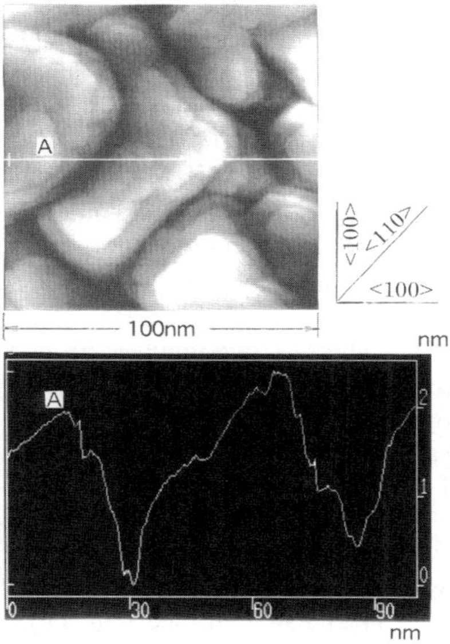

Fig. 4.18. STM images of Ag surface of MgO/Pt(10)/Cu(5)/NiFe(5)/Cu(2.8)/Co(5)/ Cu(1.2)/Ag(5) film

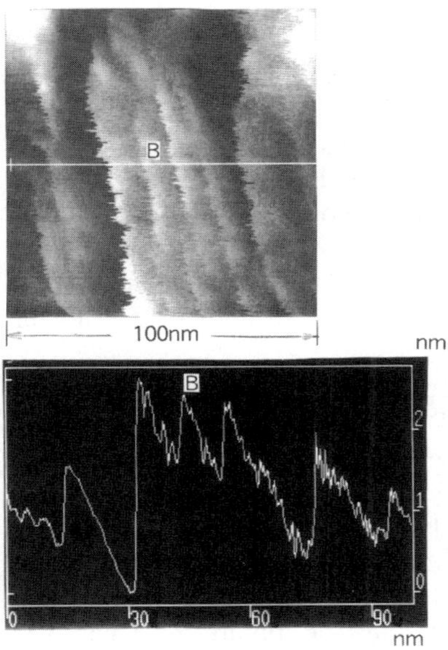

Fig. 4.19. STM images of Cu surface of MgO/Pt(10)/Cu(5)/NiFe(5)/Cu(2.8)/Co(5)/ Cu(1.2) film

The Pt(10 nm)/Cu(5 nm) layer is replaced by Cu(7 nm) for the calculation as the resistivity of Pt is about 5 times higher than that of Cu. ΔG increases with an increase of t_{Ag} due to the back-layer effect as mentioned in Sect. 4.2.1. The MR ratio increases with increase of t_{Ag} for a thin Ag capping layer due to the back-layer effect, then decreases with t_{Ag} for a thick Ag layer due to the shunting effect and takes a broad maximum at a certain value of t_{Ag}. The maximum point depends upon the relationship between the effective MFP and the film thickness. This is the reason why the MR ratio of the film with larger P takes a maximum with thinner t_{Ag}. Both the MR ratio and ΔG increase by increasing P of the capping layer due to the specular scattering effect. Comparing Fig. 4.15 with Fig. 4.17, the enhancement of the MR ratio with the addition of an Ag capping layer may be attributed to the increase of P of the capping layer. For example, the drastic increase in the MR ratio and ΔG with the addition of 1 nm thick Ag shown in Fig. 4.15 can be explained by the change in the P of the capping layer from 0 (for Cu) to 0.5 (for Ag) as shown with arrows in Fig. 4.17. The MR maximum at around t_{Ag}=2~3nm and the dependence of the MR ratio on t_{Ag} found in Fig. 4.15 are also qualitatively explained with this model assuming P~ 0.5 in Fig. 4.17.

The specularity factor of the capping layer may depend upon the surface roughness. Figures 4.18 and 4.19 shows STM images of the surfaces of the Ag and Cu capping layer of spin-valve films. The Ag capping layer is covered with square shaped plateau having smooth surfaces, with the sides parallel to the <110> direction. The roughness measured along the white line shown in the figure was less than 0.2 nm. On the other hand, Cu capping layer has a surface with large undulations. The roughness measured along the line in the figure was larger than 0.5 nm. Taking the Fermi wavelength of Cu (or Ag) into consideration, the difference in the surface roughness may contribute to the difference in the specularity between the Cu and Ag surfaces.

4.2.2.2 Specular Scattering by Oxide Pinning Layers

Large MR ratios are obtained in spin valves having oxide AF such as NiO/Co/ Cu/ Co(/Cu/Co/NiO)[10], α-Fe_2O_3/Co/Cu/Co(/Cu/Co/α-Fe_2O_3)[13], $LaFeO_3$/CoFe/ Cu/CoFe[43]. For example, NiO-based spin valves showed MR ratios of ~19% and ~24% for the single and the dual structures, and α-Fe_2O_3 based spin valves showed MR ratios of ~20% and ~28% for the single and the dual structures. These enhancements of the GMR are attributed to the specular scattering effect of the conduction electrons at the metal/insulator interfaces.

Two types of dual spin valves composed of α-Fe_2O_3 (bottom)/D-SV/ α-Fe_2O_3(top) and composed of α-Fe_2O_3(bottom)/D-SV/IrMn(top) are prepared on α-Al_2O_3(110) single crystalline substrates to study the difference in the specularity of the top AF, where D-SV represents Co(2)/Cu(2)/Co(5)/Cu(2)/Co(2). A single spin valve composed of α-Fe_2O_3(bottom)/S-SV is also prepared for comparison where S-SV represents Co(2)/Cu(2)/Co(5)/Cu(0.4). Figure 4.20 shows the MR curve for the single spin valve that showed an extremely large MR ratio of ~18% at RT owing to specular scattering at the α-Fe_2O_3/Co interface. Figures 4.21 and 4.22 show the MR curves for α-Fe_2O_3 (bottom)/D-SV/α-Fe_2O_3 (top) and for α-Fe_2O_3(bottom)/D-SV/IrMn(top). Both curves show three distinct steps: the first

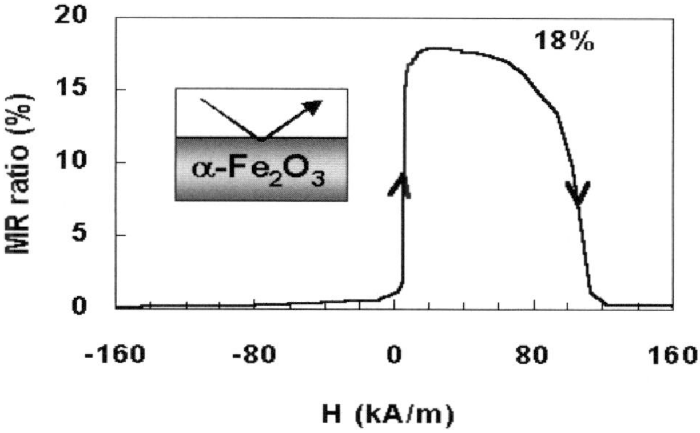

Fig. 4.20. MR curve for an α-Fe_2O_3(50)/Co(2)/Cu(2)/Co(5)/Cu(0.4) single spin valve

step is an increase of the MR ratio at the low field due to the magnetization reversal of the 5 nm-thick Co free layer. The second and third step are the first and second decrease of the MR ratio, due to the magnetization reversal of the Co pinned layers. Judging from the MR curve for the single spin valve shown in Fig. 4.20, the first and second decrease in the MR ratio correspond to the magnetization reversal of the Co pinned layer in contact with α-Fe_2O_3(top) and with α-Fe_2O_3(bottom), respectively.

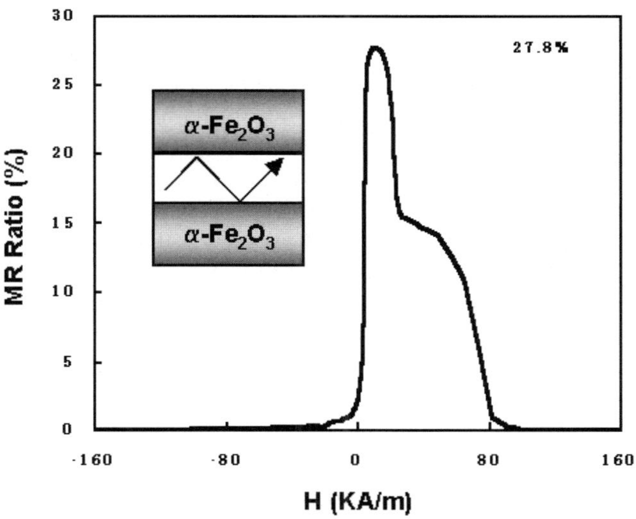

Fig. 4.21. MR curve for α-Fe_2O_3(50)/Co(2)/Cu(2)/Co(5)/Cu(2)/Co(2)/α-Fe_2O_3(50)

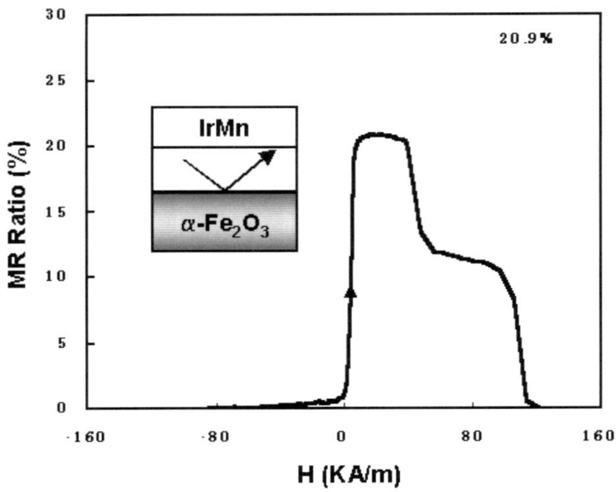

Fig. 4.22. MR curve for α-Fe_2O_3(50)/Co(2)/Cu(2)/Co(5)/Cu(2)/Co(2)/IrMn(8)

The difference in the pinning fields between the bottom and top pinned layers might originate in the difference in the crystalinity between the bottom and top α-Fe_2O_3. Figure 4.23 shows a cross-sectional TEM image of α-Fe_2O_3/D-SV/ α-Fe_2O_3 dual spin valves. The diffraction pattern showed the epitaxial growth of the α-Fe_2O_3 on the α-Al_2O_3(110) substrate. The orientational relationships are α-Al_2O_3(110)//α-Fe_2O_3(110)//Co/Cu(111) in the stacking direction and α-Al_2O_3[001]//α-Fe_2O_3[001] in the film plane. The top α-Fe_2O_3 did not show epitaxial growth and the orientation is different from that of the bottom α-Fe_2O_3. This might be the reason for the difference in the pinning fields between the top and bottom Co pinned layers. The MR ratio for the α-Fe_2O_3/D-SV/α-Fe_2O_3 dual spin valve was 27.8%, which is larger than that of 20.9% for the α-Fe_2O_3/D-SV/IrMn dual spin valve. The resistivity of the IrMn layer is much higher than that of the Co or Cu layers and the thickness of the IrMn layer is only 8 nm. Therefore, the shunting effect is considered to be small. The difference in the MR ratios between the two types of dual spin valves may originate in the difference in the specularity factor, *P*, between the Co/α-Fe_2O_3 and IrMn/vaccum interfaces. This could be understood from the TEM image shown in Fig. 4.23, where the α-Fe_2O_3(bottom)/Co interface is atomically flat owing to the epitaxial growth of bottom α-Fe_2O_3 on α-Al_2O_3, and would have a higher specularity factor than the Co/α-Fe_2O_3(top) interface that showed small undulations. Therefore, the specular scattering at the α-Fe_2O_3(bottom)/Co interface is considered to contribute the enhancement of the GMR more than that of the Co/α-Fe_2O_3(top) interface.

The MR ratio for the dual spin valve composed of Co(2)/Cu(2)/Co(5)/Cu(2)/Co(2) was calculated using Barnas's model with the parameters: λ_1=15 nm, λ_2= 60 nm, N_B=2, and $T^{\downarrow/\uparrow}$=0.9/0.1, where λ_1 and λ_2 are the MFP for the Co and Cu layers, respectively, N_B is the spin asymmetry factor, and *T* is the transmission coefficient at the Co/Cu interfaces. Figure 4.24 shows the numerical result for the dependence

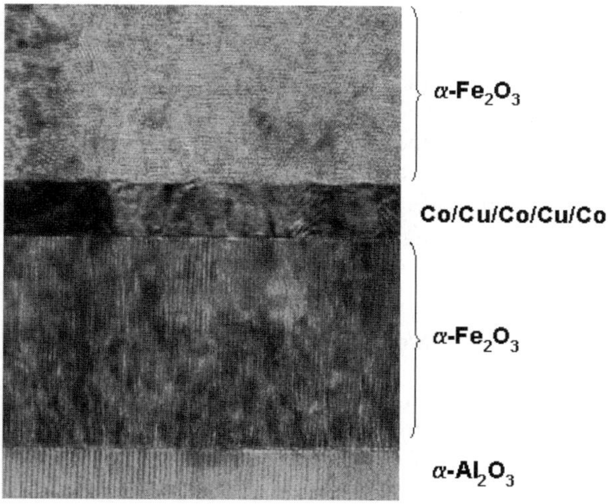

Fig. 4.23. Cross-sectional TEM image of α-Al_2O_3/α-Fe_2O_3(50)/Co(2)/Cu(2)/Co(5)/Cu(2)/Co(2)/ α-Fe_2O_3(50)

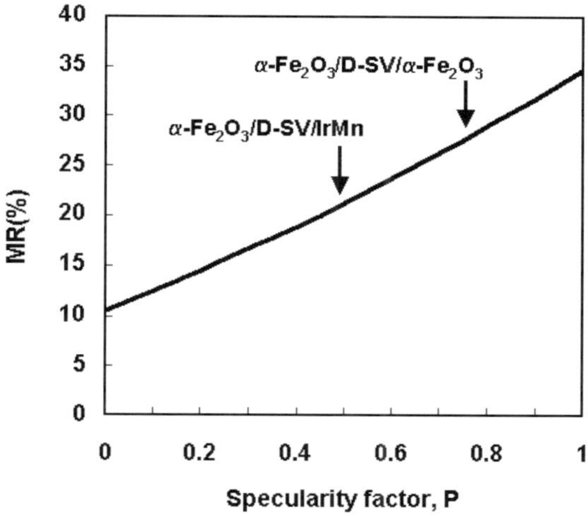

Fig. 4.24. Calculated result for the dependence of the MR ratio for Co(2)/Cu(2)/Co(5)/Cu(2)/Co(5)

of the MR ratios on the specularuty factor, *P*, for D-SV=Co(2)/Cu(2)/Co(5)/Cu(2)/Co(5). The experimental results of the MR ratios for α-Fe_2O_3/D-SV/α-Fe_2O_3 and α-Fe_2O_3/D-SV/IrMn dual spin valves are also indicated with arrows. The experimental results could be understood by assuming that the α-Fe_2O_3(bottom)/Co interface, Co/α-Fe_2O_3(top) interface, and the IrMn top surface had a specularity factor *P* of $\cong 1$, $\cong 0.5$, and $\cong 0$, respectively. Enhanced GMR by specular scattering was also found in pseudo-spin-valves having an oxide hard magnetic layer in the HM, such as HM(=$CoFe_2O_4$/CoFe)/Cu/SM/Ta[30], where HM and SM denote the hard magnetic and soft magnetic layers. CoFe ferromagnetically coupled with $CoFe_2O_4$ with a large coercivity plays the role of the HM. In this spin valve, specular scattering of the conduction electrons is expected to occur at the interface between the $CoFe_2O_4$ and CoFe. For example, $CoFe_2O_4$(50)/CoFe(2)/Cu(2.1)/CoFe(5)/Ta(3) spin-valve film showed an MR ratio of 14.8% and ΔR_S of 2.2Ω/ . The MR ratio is much higher than those of spin valves having metallic AF. Specular scattering at the $CoFe_2O_4$/CoFe interface may contribute to the enhanced GMR.

4.2.2.3 Specular Scattering by Thin Oxide Layers

Spin valves having oxide AF such as NiO and α-Fe_2O_3 show very large MR ratios owing to the specular scattering effect. But the blocking temperatures, T_B, of NiO or α-Fe_2O_3 are not sufficiently high for practical application to MR heads. On the other hand, PtMn-based spin valves have sufficiently high T_B, but show rather small MR ratios. The enhanced GMR by NOL in the PL of an IrMn-based spin valve was first reported by the Tohshiba group [41], where the NOL is a nano-

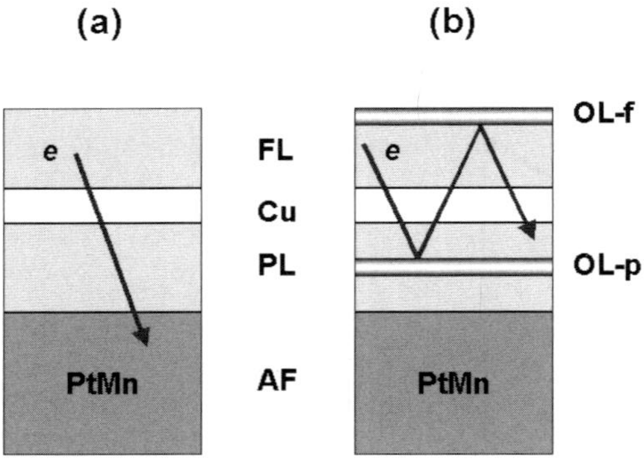

Fig. 4.25. Electron trajectories in PtMn-based spin valves **(a)** without and **(b)** with OL-p and OL-f

meter oxide layer prepared by the natural oxidation of the PL(=CoFe). The spin valve has a structure of Ta(5)/NiFe(2)/IrMn(7)/PL(with NOL)/Cu(2)/ CoFe(2)/ CoFeO(0.4)/TaO(0.4) and showed a large MR ratio of 16%, with a coercivity of the free layer, 14 Oe. Then, specular scattering by a thin oxide layer inserted in PtMn-based spin valves was also studied [42], as the PtMn-based spin valves show better thermal stability than the IrMn-based ones. Figure 4.25 shows an illustration of the electron trajectories for the parallel magnetization alignment in a conventional PtMn-based spin valve and that with OL-p and OL-f, where OL-p and OL-f denote thin oxide layers inserted in the PL and FL. The electron MFP in the PtMn layer is very small, which reduces the effective MFP in conventional PtMn-based spin valves, resulting in the small MR ratios as shown in Fig. 4.25(a). If the interfaces of PL/OL-p and FL/OL-f shown in the Fig. 4.25(b) are specularly reflective, a large enhancement of the MR ratio is expected.

Firstly, let us consider the effect of OL-p inserted in the PL of the spin valves composed of Ta(3)/PtMn(15)/CoFe(4)/Cu(2.2)/CoFe(5)/Ta(3) having OL-p in the PL. Figure 26(a) and (b) show the MR curve without OL-p and that with OL-p (0.6). Large increases in the MR ratio from 8.1% to 11.2% and in ΔR_s from 1.17Ω to 1.62Ω were observed with the insertion of only 0.6nm thick OL-p. Figure 4.27 shows the dependence of the MR ratio, R_s, and ΔR_s on the OL-p thickness, t, for spin valves. The MR ratio and ΔR_s increased drastically with the insertion of OL-p for the OL-p thickness, t 0.6 nm, then the values slightly increased with t >0.6 nm and saturated for t 1.4 nm. On the contrary, R_s showed no remarkable changes with the increase of t. The drastic increases in MR ratios and in ΔR_s with the insertion of OL-p may originate in the specular scattering of the conduction electrons by OL-p, though it still remains as a matter of debate.

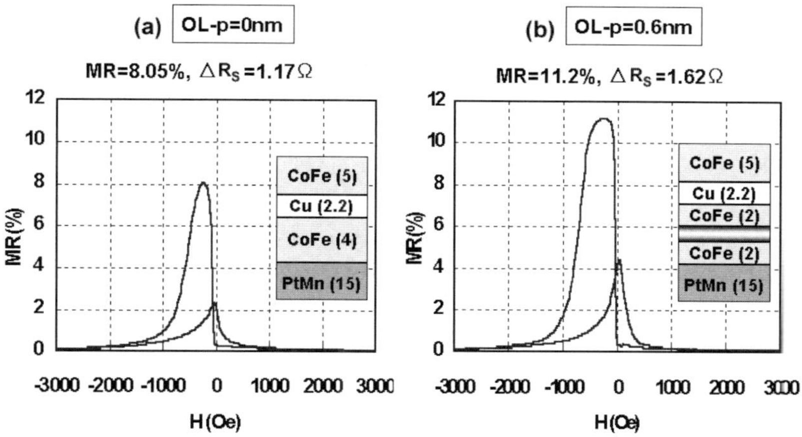

Fig. 4.26. MR curves for the spin valve **(a)** without OL-p and **(b)** with OL-p(0.6 nm)

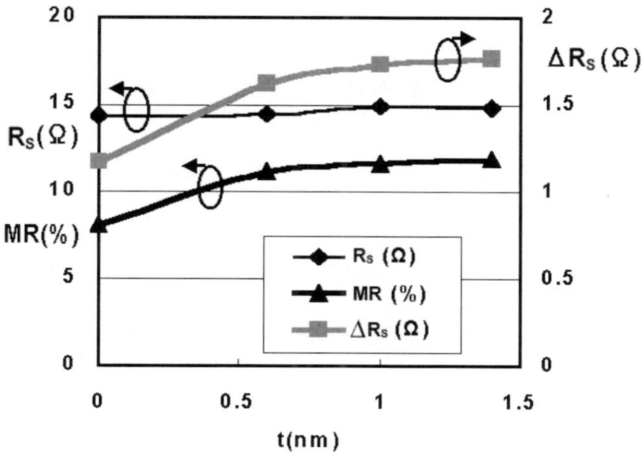

Fig. 4.27. Dependence of the MR ratio, R_s, and ΔR_s on the OL-p thickness, t for the spin valves composed of Ta/PtMn/CoFe/Ol-p(t)/CoFe/Cu/CoFe/Ta

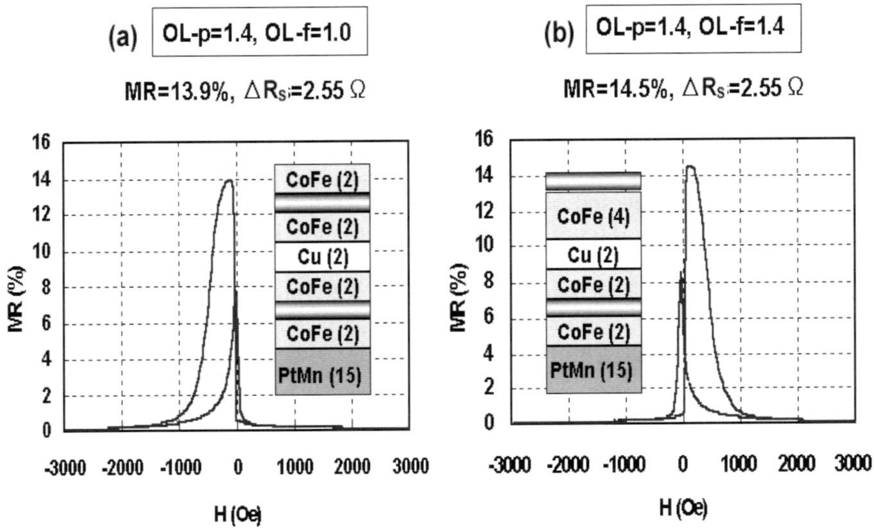

Fig. 4.28. MR curves for spin valves having (**a**) OL-f in the FL and (**b**) OL-f on the FL

Let us consider the MR properties of spin valves having both the OL-p and OL-f. Figure 4.28(a) shows MR curve for a spin valve having an OL-p(1.4) in the PL and an OL-f(1.4) on the FL. The MR ratio and ΔR_S were 14.5% and 2.55Ω, respectively. Figure 4.28(b) shows the MR curve for the spin valve having an OL-p(1.4) in the PL and an OL-f in the FL. The MR ratio and ΔR_S were 13.9% and 2.55Ω, respectively. Both spin valves showed larger MR ratios than the spin valve having only OL-p as shown in Fig. 4.26. Therefore, both OL-p and OL-f are considered to contribute to the enhancement of the GMR. These results can be explained by the specular scattering of the conduction electrons at the PL/OL-p

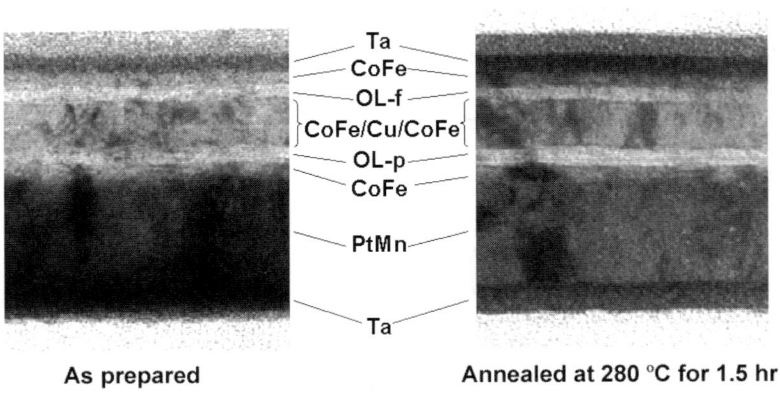

Fig. 4.29. Cross-sectional TEM images of the Ta/PtMn/CoFe/OL-p/CoFe/Cu/CoFe/OL-f/CoFe/Ta spin valve before and after annealing

interface and at the FL/OL-f interface as illustrated in Fig. 4.25. It is desirable for the OL-p and OL-f to have smooth surfaces to cause specular scattering. Figure 4.29 shows a cross-sectional TEM image of the spin valve shown in Fig. 4.28(b), before and after annealing at 280 °C for 1.5 hr to obtain the ordered state of PtMn. It is interesting to note that both OL-p and OL-f have smooth surfaces only after annealing, but not when they are prepared. The large enhancement of MR ratios and ΔR_s by the insertion of OL-p and OL-f may be explained by assuming that both OL-p and OL-f contribute to the specular scattering of the conduction electrons in PtMn-based spin valves after the annealing done for the ordering of PtMn. The specurality factor, *P*, for the spin valve shown in Fig. 4.28(a) was estimated using Barnas's model [36], taking account of the shunting effect of Ta and PtMn. The estimated value of *P* was 0.6~0.7. So, further enhancement of the GMR could be obtained by improving the specularity of the interfaces between the PL/OL-p and OL-f/FL.

The SAF spin valves composed of AF/SAF(=PL1/Ru/PL2)/Cu/FL show better thermal stability than the simple spin valve composed of AF/PL/Cu/FL as mentioned in Sect. 4.1.1. However, the MFP in the Ru layer in the SAF is very short, resulting in the reduction of the effective MFP in the SAF spin valves. Therefor, the OL-p for the SAF spin valves should be inserted in the PL2, to enhance the MR ratios, though this makes the spin valves have a complicated strucure of AF/PL1/Ru/PL2′/OL-p/PL2/Cu/FL. Figure 4.30 shows the MR curves of the SAF spin valve (a) having only the OL-p in the PL2 and (b) having the OL-p in the PL2 and OL-f on the FL. The MR ratio, ΔR_s, and R_s were 10.7%, 2.58Ω, and 24.1Ω, respectively for spin valve (a) and were 13.0%, 2.62Ω, and 20.2Ω, respectively for spin valve (b). The SAF spin valves with OL-p and OL-f showed smaller MR ratios than the simple spin valves with OL-p and OL-f,

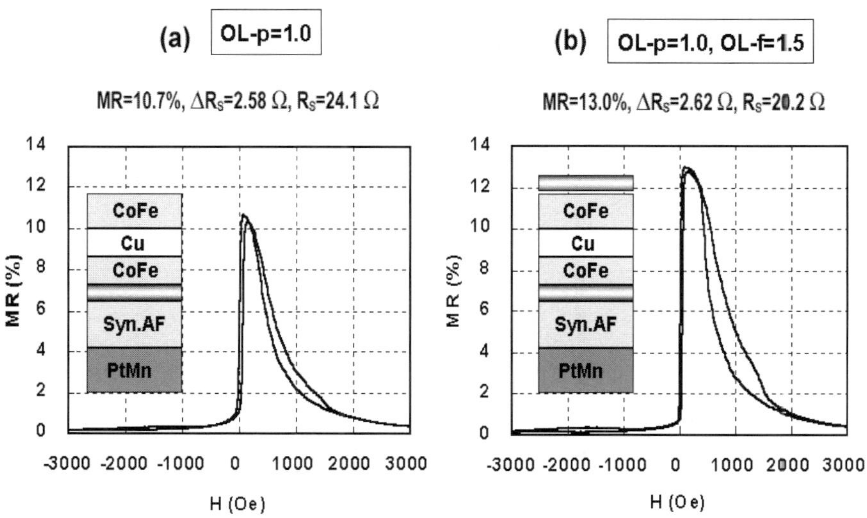

Fig. 4.30. MR curves for the SAF spin valves **(a)** having OL-p and **(b)** having both OL-p and OL-f

probably due to the shunting effect of the AF/PL1/Ru/PL2′. Attention should be paid to the difference in R_S between the SAF spin valve without OL-f and with OL-f in Fig. 4.30. A drastic decrease in R_S, from 24.1Ω/□ to 20.2Ω/□, was observed with the addition of the OL-f. This result might be the proof of the specular scattering effect caused by the addition of the OL-f.

Dual-structured spin valves having specular reflective oxide layers in the pinned layers were also studied for PtMn and IrMn based ones [42, 44]. For example, Ta(3)/PtMn(15)/CoFe(2)/OL-p(1)/CoFe(2)/Cu(2.2)/CoFe(5)/Cu(2.2)/CoFe(2)/OL-p(1)/CoFe(2)/PtMn(15)/Ta(3) showed a large MR ratio of 17.3% owing to specular scattering at the interfaces of the top and bottom OL-p/CoFe. But R_S was reduced to 1.3 Ω because of the small R_S of the dual spin-valve film. A very large MR ratio of 23.6% was reported for the dual spin valve composed of Ta(3)/NiFe(2)/IrMn(5)/NOL/CoFe(2)/Cu(2)/CoFe(2)/Cu(2)/CoFe(2)/NOL/IrMn(5)/Ta(3). In this case, specular scattering occurs at the interfaces of the top and bottom NOL/CoFe. A PdPtMn-based spin valve having specular reflective oxide layers composed of Ta/NiFe/PdPtMn/CoFeB/Ru/CoFeB/OL/CoFeB/Cu/CoFeB/FeO was studied and the film showed a large MR ratio of 15.7% and a large R_S of 3.8Ω [45]. The specular scattering occurs at the interfaces of OL/CoFeB and CoFeB/FeO in this case. Addition of OL-f to the α-Fe_2O_3 based spin valves was also studied. Figure 4.31 shows the MR curves for α-Fe_2O_3(50)/Co(2)/Cu(2.2)/Co(2)/Ta(3) single spin valves prepared on an Al_2O_3 substrate with and without OL-f(1.5) inserted between the Co(2) and Ta(3). The addition of OL-f increased the MR ratio and ΔR_S from 17.3% to 20.9% and from 7.01Ω/□ to 7.85Ω/□, respectively and decreased R_S from 40.4Ω/□ to 36.9Ω/□. The result is well explained by assuming that the specular scattering occurs at the interfaces of α-Fe_2O_3/Co and Co/OL-f in this case. Similar results were obtained for $CoFe_2O_4$-based pseudo-spin-valve films. For example, $CoFe_2O_4$(50)/CoFe(2)/Cu(2.1)/CoFe(2)/OL(1)/CoFe(2)/Ta(3) showed an MR ratio of 18% and ΔR_S of 3.4Ω/□; $CoFe_2O_4$(30)/CoFe(2)/Cu(2.1)/CoFe(2)/OL(1.4)/Ta(3) showed an MR ratio of 17.3% and ΔR_S of 3.8Ω/□. These spin valves have a center portion

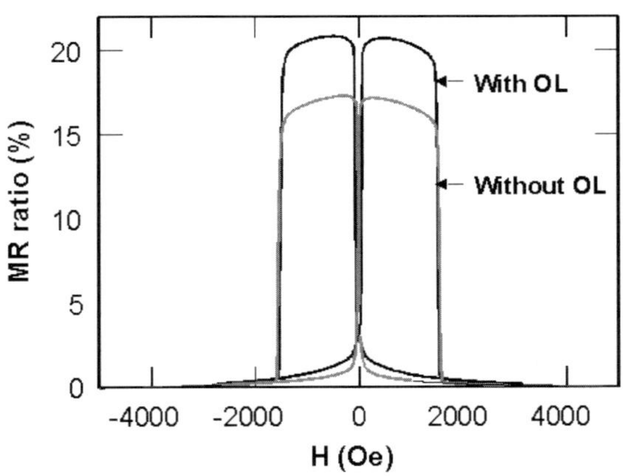

Fig. 4.31. MR curves for α-Fe_2O_3 based spin-valves with and without OL-f

composed of very thin metallic layers, CoFe(2)/Cu(2.1)/CoFe(2), bounded with oxide layers. This is the reason why these spin valves showed very large ΔR_s and large MR ratios owing to the specular scattering effect at the interfaces of $CoFe_2O_4$/CoFe and CoFe/OL. Various types of spin valves having specular reflective layers are studied for the application to MR heads. The details will be described in Sect. 4.3.

4.3 Magnetic Read Heads

Research and development on GMR have been done vigorously owing to the suitability for read heads used for HDD. Since the launching of MR heads using AMR films at the beginning of the 1990s, the yearly growth of the areal bit density in HDD increased from 30% to 60%. The growth rate has now accelerated from 60% to 100% by the introduction of GMR heads using spin valves in 1998 as shown in Fig. 4.32. According to the trend, an area density of 100 Gb/inch^2 could be achieved before 2005 and 3.5 inch HDD with 100 GB or more will become commercial in the near future. These drastic increases in the areal density and storage capacity of the HDD extended the business fields of HDD, not only in PC but also in AV (audio-visual) applications. For example, 20 hour video recording in MPEG2 format with 4 Mbps is possible by using a 36 GB HDD. Ultra-high-capacity HDD will replace the VTR in future, having the advantage of random access operation and suitability for PCs. Another challenge is a downsizing of the HDD to memory card size, such as a "Microdrive" loaded with 1 inch HDD of 1 GB developed by IBM. The memory card sized HDD might be a competitor to semiconductor memory cards and will be used for digital cameras, PDA, and

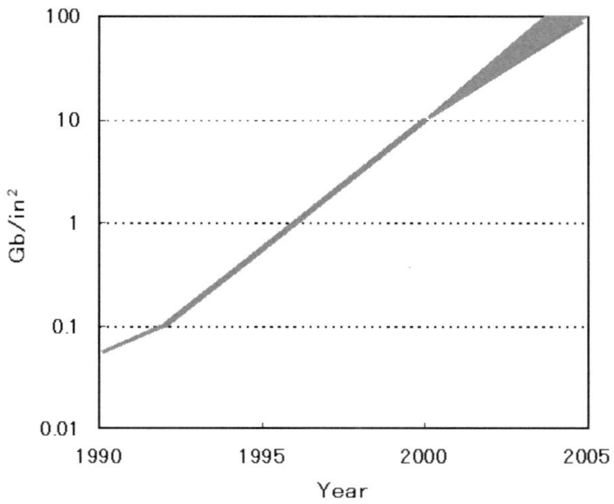

Fig. 4.32. Areal recording density growth in HDD

portable audio and video recorders. GMR is a key device for ultra-high-density storage devices up to 100 Gb/inch^2. TMR or CPP (current perpendicular to the plane) MR would be the candidate beyond 100 Gb/inch^2.

4.3.1 Head Design

Inductive heads were used for HDD until the introduction of MR heads at the beginning of 1990s. The output of the inductive heads for the reproducing operation is obtained by the flux change with time, $\partial\phi/\partial t$. Therefore, the output is proportional to the bit size and the angular velocity, $r\omega$, of the disk. Development in HDD storage devices made the bit size and the disk diameter smaller and smaller, resulting in a decrease of the output of the inductive heads. The output of the MR heads does not depend upon the angular velocity, which is advantageous for HDD having disks with smaller diameters. Furthermore, MR heads show higher output than inductive heads, simply increasing the MR ratios. This is the main reason why MR heads replaced the inductive heads according to the development of high-density recording in the 1990s. Figure 4.33 shows a typical structure of an MR head having a write head portion and a read head portion. The write head portion has a write pole composed of high B_s materials and a write gap for signal writing using the magnetic flux generated by the write coils. The structure is the same as that of the inductive head in principle. The read head portion has a (lower) shield, a shield gap, in which the MR element is placed insulated from the lower shield and the shared pole which works as a lower pole for the read head portion and as an upper shield for the read head portion.

Figure 4.34 shows a typical structure of the MR element in a shield gap, named "abutted structure". The MR element is composed of a spin-valve film, SV, with a

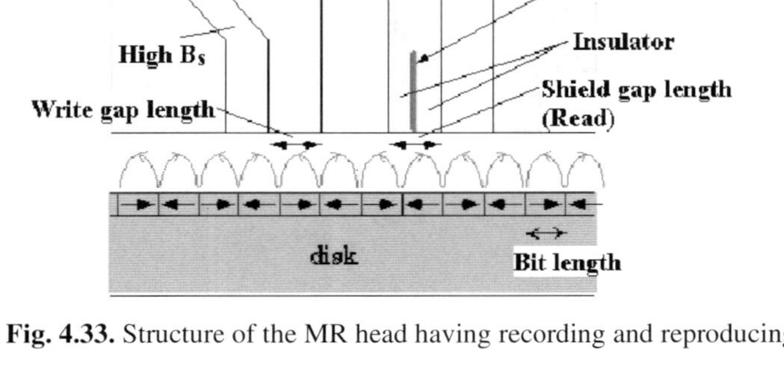

Fig. 4.33. Structure of the MR head having recording and reproducing portion

cap, CL, and an underlayer, UL, leads, L, and hard magnets, HM, that stabilize the single domain structure of the free layer, FL, of the SV. The easy axis of the FL makes a right angle against that of the pinned layer, PL. The direction of the signal field is parallel to the easy axis of the PL and perpendicular to that of the FL so that only the magnetization rotation of FL occurs with the signal field to avoid Barkhausen noise. The width of the SV between the HM works as the track width, T_W, of the read head. The width of the MR element along the gap depth direction is called the "MR height" and is indicated as h in Fig. 4.34. The track width becomes shorter year by year with the development of high-density recording. This makes the patterning process of the SV element more difficult and decreases the active area of the FL in the T_W region.

A new type of structure named an "overlaid structure" has been developed to cope with the narrower gap structure as shown in Fig. 4.35. In this case, the width between the leads works as the T_W of the read head. So the size of the SV could be larger than the T_W, which makes the patternng process of the SV element easier and increases the active area of the FL in the Tw region. In both types of MR elements, the sense current flows parallel to the easy axis of the FL and generates a bias field, H_I, to the FL. The stray field from the PL also acts as the bias field,

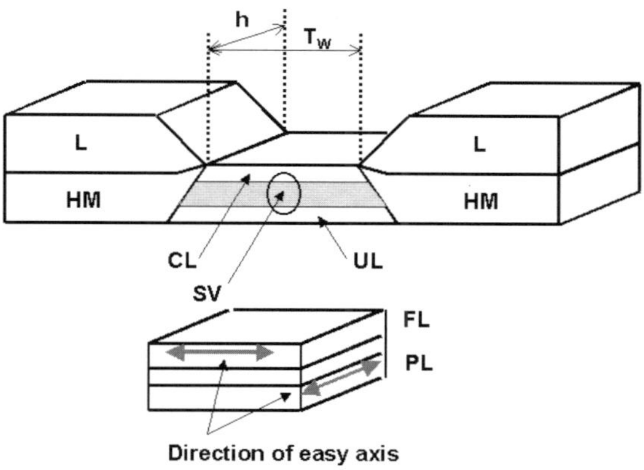

Fig. 4.34. MR element in the read head having an abutted structure

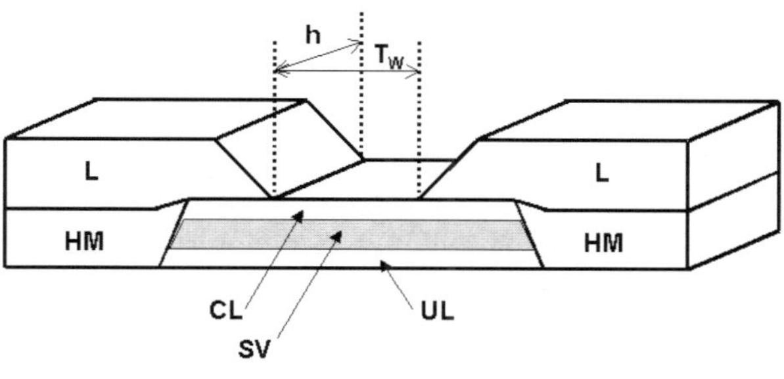

Fig. 4.35. MR element in the read head having an overlaid structure

H_M, for the FL. This effect becomes larger when the ratio of the patterned size to the film thickness of the PL becomes smaller. Therefore, it is necessary to balance the H_I and H_M to make the bias point of the field close to zero. Spin valves having a SAF structured PL enable one to reduce the H_M. Spin filter spin valves have a high conduction layer, HC, in contact with the FL, which causes a bias field, H_F. In this case, it is necessary to balance H_I, H_M, and H_F. But this is not difficult, as the H_M is usually small and the direction of the H_F is opposite to that of the H_I. The HC in the spin filter SV plays the very important role of balancing the bias fields, not just to enhance the MR ratio as mentioned in Sect.4.2.1.

The output of the spin-valve heads is given as

$$\left.\begin{aligned} \Delta V &\propto I\cdot \Delta R\cdot \Phi/(M_F\cdot t_F) \\ &= I\cdot(\Delta\rho/d)\cdot(T_W/T_H)\cdot \Phi/(M_F\cdot t_F) \\ &= I\cdot \Delta R_S\cdot(T_W/T_H)\cdot \Phi/(M_F\cdot t_F) \end{aligned}\right\} \quad (4.15)$$

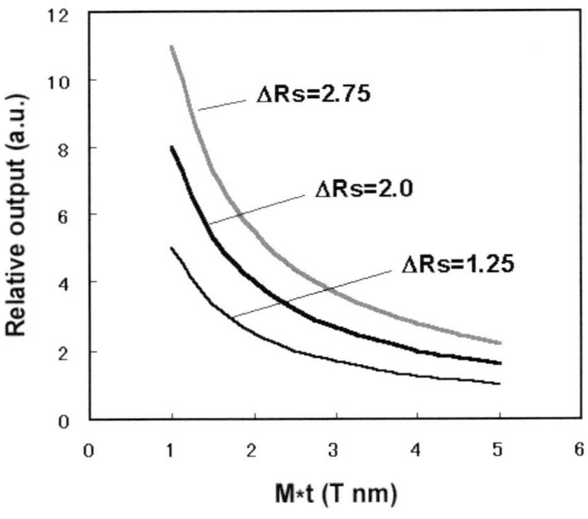

Fig. 4.36. Dependence of the relative output on the $M_F\cdot t_F$ and ΔR_S

where T_W and T_H are the track width and the MR height of the MR element, respectively. Φ is the flux from the media. M_F and t_F are the saturation magnetization and the thickness of the FL of the spin-valve film, respectively. The ratio T_W/T_H is almost constant. Therefore, higher output is obtained when the spin valve has a larger ΔR_S and a smaller $M_F\cdot t_F$. Figure 4.36 shows the dependence of the relative output on the $M_F\cdot t_F$ and ΔR_S calculated using (4.15) with the condition that $(T_W/T_H)\cdot\Phi$ is constant. The output is normalized with that for ΔR_S=1.25 (Ω) and $M_F\cdot t_F$ =5 (Tnm).

The reduction of $M_F \cdot t_F$ is very effective in increasing the output. But the simple reduction of t_F causes deterioration of the soft magnetic property of the FL and a decrease of the MR ratio. Spin-filter spin valves [35] make it possible tc reduce t_F and enhance the MR ratio without degrading the soft magnetic property of the FL as mentioned in Sect. 4.2.1. M_F can be reduced by using a Ni-rich magnetic layer for the FL. But MR ratios of spin valves having Ni-rich magnetic layers are smaller than those of spin valves having Co-rich magnetic layers. Therefore, it is not easy to reduce M_F without degrading the MR ratio. This problem could be solved by using a SAF structured FL [46, 47]. Figure 4.37 shows structure of the spin valve that has a SAF free layer and SAF pinned layer, where the free layer is composed of the first and the second magnetic layer, FL1 and FL2, antiferromagnetically coupled to each other across the nonmagnetic layer(Ru) and the pinned layer is composed of antiferromagnetically coupled PL1 and PL2 across the nonmagnetic layer (Ru). The effective $M_F \cdot t_F$ of the free layer is given as

$$M_F \cdot t_F = | M_1 \cdot t_1 - M_2 \cdot t_2 | \tag{4.16}$$

where M_i and t_i (i=1 or 2) are the magnetization and thickness of the first and second magnetic layer in the SAF structures FL. Then, the effective $M_F \cdot t_F$ can be reduced to a small value by adjusting the value of $M_1 \cdot t_1$ close to that of $M_2 \cdot t_2$. The magnetization directions of M_1 and M_2 in the SAF structured FL rotate with the application of a small magnetic field, H, keeping the antiparallel alignment. For example, if $M_2 \cdot t_2$ is larger than $M_1 \cdot t_1$, M_2 rotates in the direction of the applied field and M_1 rotates in the opposite direction as shown in Fig. 4.37. The driving force for the rotation of FL comes from the Zeeman term, $H \cdot (M_1 \cdot t_1 - M_2 \cdot t_2)$. Therefore, $M_1 \cdot t_1$ should not be equal to $M_2 \cdot t_2$ in the SAF structured FL, while this is not necessary for the SAF structured PL.

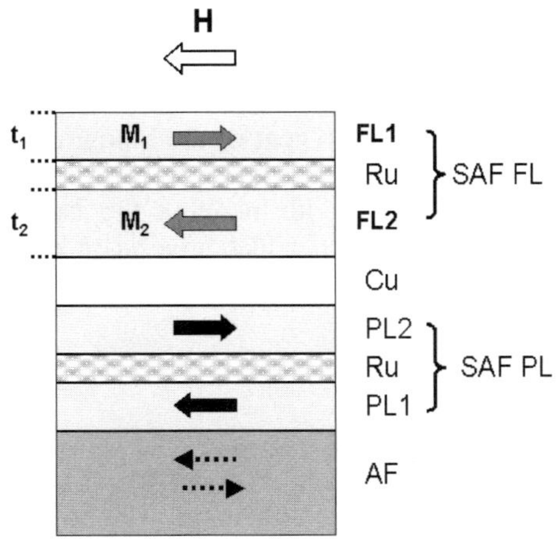

Fig. 4.37. Structure of the spin valve having a SAF structured FL

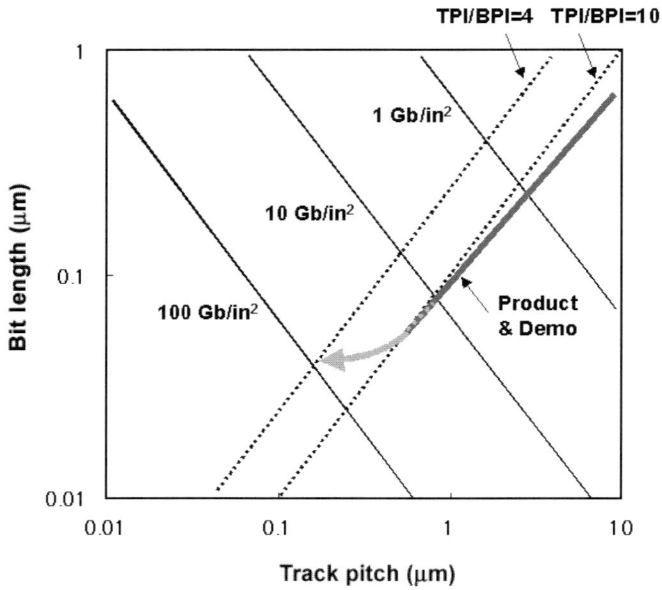

Fig. 4.38. Trend of recording bit sizes in HDD

Figure 4.38 shows the trend of recording bit sizes in HDD and Table 4.3 shows an example of the recording bit sizes for 10, 40 and 100 Gb/in^2. The aspect ratio, track pitch against bit length, decreases with decrease of the bit size and is expected to be 4 or so for ultra-high-density recording over 100 Gb/inch^2. It is required to reduce the shield gap length to cope with the small bit length. But the minimum thickness of the MR element and that of the insulator films placed between the MR element and the lower and upper shields as shown in Fig. 4.33, are ~30 nm and ~25 nm at present, respectively. Therefore it is very difficult to make the shield gap length narrower than 80 nm. On the other hand, the track width has been decreased drastically owing to the development of the lithography technique. This is the main reason for the decrease in the aspect ratio for higher recording density. However, the reduction of the track width increases the aspect ratio of the write pole thickness against the track width, which gives rise to a serious difficulty in the write pole manufacturing process at present.

Table 4.3. Example of the bit size for 10, 40, and 100 Gb/in^2

	10Gb/in^2	40Gb/in^2	100Gb/in^2
Kbpi	330	570	630
Ktpi	30	70	160
Track pitch	0.83 μm	0.36 μm	0.16 μm
Bit length	75 nm	44 nm	40 nm
Aspect ratio	~ 11	~ 8	~ 4

The decrease in the bit size with the development of high-density recording makes the size of the MR element smaller and smaller, resulting in an increase of the current density of the MR element. In (4.15), we assume constant current operation, but this will not stand for the small sized MR head, because of the upper limit of the current density. Therefore, constant current density operation will be required for future read heads. Then (4.15) is replaced by

$$\left.\begin{aligned} \Delta V &\propto I\cdot\Delta\rho\cdot T_W/(d\cdot T_H)\cdot\Phi/(M_F\cdot t_F) \\ &= J\cdot\Delta\rho\cdot T_W\cdot\Phi/(M_F\cdot t_F) \\ J &= I/(d\cdot T_H) \end{aligned}\right\} \quad (4.17)$$

where J is the current density of the MR element. In this case, $\Delta\rho$ and T_W becomes the important factors, while ΔR_S is one of the important factors in the case of the constant current condition. Let us consider the changes in the output, when the track width and the MR height are reduced to 1/5, and the bit length is reduced to 1/2. In the case of the constant-current condition, (4.15) becomes

$$\Delta V' \propto I\cdot\Delta R_S\cdot (1/2)/(M_F\cdot t_F) \quad (4.15')$$

and in the case of the constant current-density condition, (4.17) becomes

$$\Delta V'' \propto J\cdot\Delta\rho\cdot(1/5)\cdot(1/2)/(M_F\cdot t_F). \quad (4.17')$$

Here, we assume that Φ decreases in proportion to the reduction of the bit length. It turns out that a 10 times increase in $\Delta\rho/M_F\cdot t_F$ for the constant current-density condition and a 2 times increase in $\Delta R_S/M_F\cdot t_F$ for the constant current condition are required to achieve a 10 times increase in the recording bit density. Therefore it is more difficult for the constant current-density condition than for the constant current condition to cope with the reduction of bit size. In both cases, the reduction of $M_F\cdot t_F$ and the increase of the MR ratio are key points to increase the output of the read heads. The spin filter structure or the SAF structured FL would be effective to decrease $M_F\cdot t_F$ and the specular reflective structure would be effective to increase the MR ratio.

There are many challenges to achieve high-density recording over 100 Gb/in^2. TMR (tunneling magnetoresistance) and CPP (current perpendicular to the plane) MR [48] are two candidates for future MR elements. The sense current flows perpendicular to the film plane in these elements by using the upper and lower shield as the leads. This structure enables the read heads to have narrower shield gaps, because a TMR or CPPMR head does not need the insulative films between the element and the shields, unlike the GMR head. The TMR head (see Sect .5.2) enables a large output, but the RA (resistance area product) is larger than 10 Ω·μm at present. This value is too large for a read head having a track width narrower than 0.3μm. On the contrary, The CPPMR element is suitable for a read head having a narrower track width, but the resistivity is too small at present to obtain sufficient output.

4.3.2 Spin Valves for Read Heads

Various types of spin valves have been developed for GMR heads. Among them, PtMn-based spin valves show excellent thermal stability of the MR properties owing to the high T_B > 350 °C and are widely used for read heads. A typical example of PtMn-based simple spin valves is Ta/PtMn/Co(Fe)/Cu/Co(Fe)/NiFe/Ta. The MR ratio and ΔR_S are about 7~8% and 1.2~1.3 Ω/□, respectively. The exchange bias fields, H_{ex}, of simple spin valves is about 800 Oe and is increased by using a SAF structured PL up to 2.5 kOe. A typical example of spin valves having a SAF pinned layer is Ta(3)/PtMn(15)/Co(Fe)(3)/Ru(0.8)/CoFe(2)/Cu(2.2)/Co(Fe)(1)/NiFe(4)/Ta(3) [49]. The MR ratio and ΔR_S are about 7~8% and 1.3 Ω/□. The MR ratio increases by using a dual spin valve structure. For example, the SAF dual spin valve Ta(3)/PtMn(15)/Co(3)/Ru(0.8)/Co(2)/Cu(2.2)/Co(1)/NiFe(3)/Co(1)/Cu(2.2)/Co(2.5)/Ru(0.8)Co(2)/PtMn(15)/Ta(3) showed an MR ratio of 11% and PtMn(13)/Co(t_1)/Ru(0.8)/Co(t_2)/Cu(2.2)/Co/Cu(2.2)/Co(t_2)/Ru(0.8)/Co(t_1)/PtMn(13)/Ta(10) (t_1=1.2~1.4, t_2=1.7~2.2) showed an MR ratio of about 13~14% and a ΔR_S of 1.9 Ω/□ [21].

The sensitivity of spin valves increases by decreasing the effective $M_F \cdot t_F$ of the FL using the SAF structure as shown in Fig. 4.37. Figure 4.39 shows the MR curve for a spin valve having SAF structured FL and PL; Ta(3)/PtMn/Co(2)/Ru(0.8)/Co(3)/Cu(2.2)/Co(0.5)/NiFe(3.5)/Ru(0.8)/NiFe(2)/Ta(3) [47]. It is well understood from the MR curve shown in Fig. 4.39 that the SAF structured FL behaves like one FL without deterioration of the soft magnetic property. Figure 4.40 shows the dependence of the MR ratio on the effective FL thickness, t_{FL}, for PtMn-based spin valves having SAF structured PL and FL composed of PtMn(9.3)/Co(2)/Ru(0.8)/Co(3)/Cu(2.2)/Co(0.5)/NiFe(3)/Ru(0.8)/NiFe(t)/Ta(3), the spin-filter spin valve composed of PtMn(9.3)/Co(1.5)/Ru(0.8)/Co(2.5)/Cu(2.2)/Co(0.3)/NiFe(t)/Cu(1.5)/Ta(3), and the conventional SAF spin valve composed of PtMn (9.3)/Co(1.5)/Ru(0.8)/Co(2.5)/Cu(2.2)/Co(0.3)/NiFe(t)/Ta(3). The effective thickness of the SAF structured FL is defined as (3-t) nm for a spin valve having a SAF structured FL. The MR ratio starts to decrease at around t_{FL} ~ 3 nm for the conventional spin valve having SAF structured PL, and at around 2 nm for the spin-filter spin valve, but not for the spin valve having SAF structured FL and PL. The small decrease in the MR ratio with increase of t_{FL} in the spin valve having SAF structured FL is due to the shunting effect. This SAF structured FL enables a drastic reduction of the effective $M_F \cdot t_F$ without degrading the MR ratio and might contribute to a large increase in the output of the read head.

The enhancement of the MR ratios in PtMn-based spin valves was achieved by using a specular reflective thin oxide layer, OL-p inserted in the PL and/or OL-f on the FL as mentioned in Sect. 4.2.2. For example, SAF structured PtMn-based spin valves composed of Ta/PtMn/CoFe/Ru/CoFe/OL-p/CoFe/Cu/CoFe/NiFe/OL-f/Ta showed an MR ratio of 12.3% and a ΔRs of 2.6 Ω/□ [50]. Specular scattering of the conduction electrons occurs at the interface of OL-p/CoFe and NiFe/OL-f. The OL-p is composed of oxide magnets so that the CoFe/OL-p/CoFe behaves as one magnetic layer, ferromagnetically coupled each other. Figure 4.41 shows the temperature dependence of H_{ex} for PtMn-based SAF spin valves with

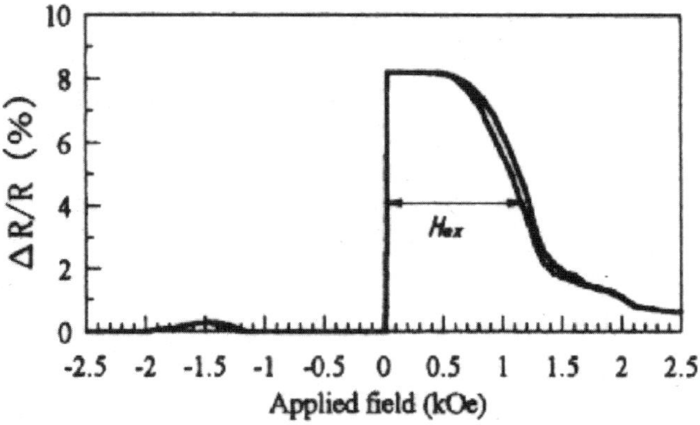

Fig. 4.39. MR curve for the spin valve having SAF structured FL and PL [47]

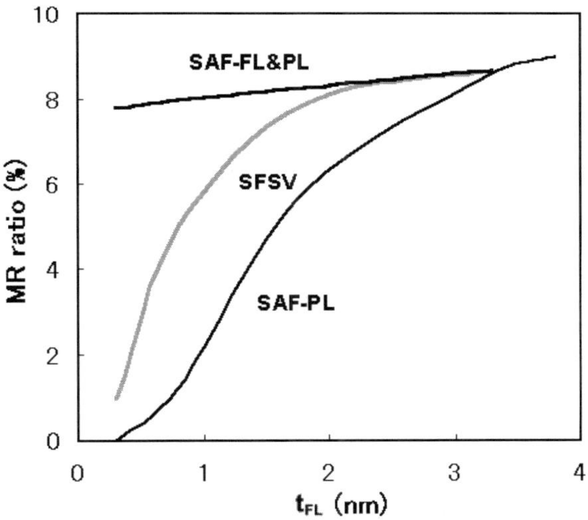

Fig. 4.40. Dependence of the MR ratio on the effective FL thickness, t_{FL}, for the spin valve having SAF structured FL and PL, spin-filter spin valve (SFSV), and conventional spin valve having SAF structured PL

and without OL. PtMn-based spin valve with OL-p and OL-f showed very high T_B and excellent thermal stability of H_{ex}, like the conventional PtMn-based SAF spin valve. The T_B of these spin valves can be increased using a Ni-based underlayer up to 400 °C [51].

PdPtMn-based spin valves also show high thermal stability. Figure 4.42 shows the temperature dependence of H_{ua} for exchange coupled AF/NiFe films [52~54]. The T_B for PdPtMn is slightly lower than that for PtMn, but higher than those for IrMn, NiO and FeMn. MR properties of PdPtMn-based spin valves are almost the same as those of PtMn-based spin vlaves. For example, the simple spin valve Ta/FL(3.5)/Cu(2.8)/PL(2.2)/PdPtMn(15)/Ta shows an MR ratio of 8% and a ΔRs of 1.3Ω/□ and the SAF spin valve Ta(5)/FL(3.5)/Cu(2.8)/PL2(2)/Ru (0.8)/PL1(1)/PdPtMn(10)/Ta(6) shows an MR ratio of 7.2% and a ΔRs of 1.5Ω/□. Specular reflective PdPtMn-based spin valves were also studied. For example, SAF spin valves composed of UL/CoFeB/Ru/CoFeB/Cu/CoFeB/CoFeBO/CuO/Al_2O_3 shows MR ratios of 10~11% and ΔRs of 2.16~2.52Ω/□ [55], respectively and the demonstration of 56 Gb/in^2 was done with this spin valve by the Fujitsu group

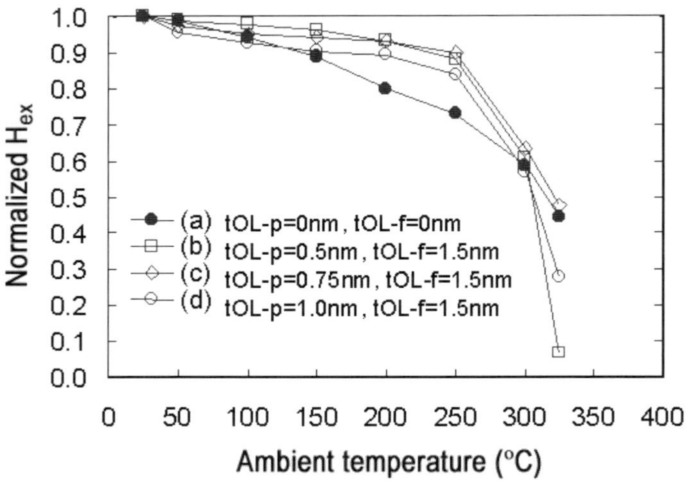

Fig. 4.41. Temperature dependence of H_{ex} for PtMn-based spin valves composed of Ta/PtMn/CoFe/Ru/CoFe/(OL-p)/CoFe/Cu/CoFe/NiFe/(OL-f)/Ta

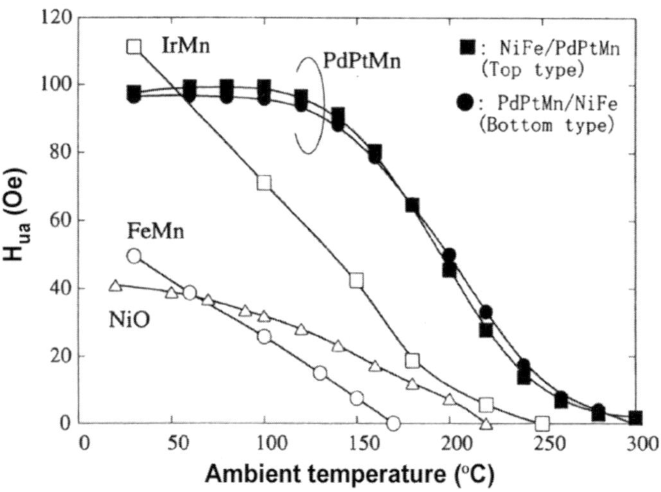

Fig. 4.42. Temperature dependence of H_{ua} for exchange coupled AF/NiFe films [52]

[56]. It was reported that the simple oxidation of CoFeB decreased the MR ratio and caused a positive shift of H_{in}, but both the oxidation of CoFeB and Cu increased the MR ratio and caused a negative shift of H_{in}.

The IrMn-based spin valves show larger MR ratios than (Pd)PtMn-based spin valves and the spin valves having specular reflective NOL [41] show large MR ratios as described in Sect. 4.2.2. The spin-filter spin valves [35], such as Ta/HCL/CoFe/Cu/CoFe/Ru/CoFe/IrMn, showed excellent MR properties, such as MR~10%, and ΔRs~2Ω/□, for the read heads as shown in Table 4.2 in Sect. 4.2.1. The critical (or minimum) thickness of IrMn required for the spin-valve device is very thin compared with other antiferromagnets as shown in Table 4.1 in Sect. 4.1.1. Therefore, the IrMn-based spin valves are suitable for dual structured spin valves. For example, the dual spin valve composed of Ta/UL/IrMn/CoFe/Ru/CoFe/Cu/CoFe/NiFe/CoFe/Cu/CoFe/Ru/CoFe/IrMn/Ta showed an MR ratio of 11.1% and a ΔRs of 1.23Ω/□ [20]. The total thickness of IrMn-based spin valves is thinner than those of other spin valves using metallic antiferromagnets such as (PdPtMn), NiMn and FeMn, which enables one to cope with the read heads having narrow shield gaps. The only demerit of IrMn-based spin valves is in the thermal stability due to the low T_B and the higher concentration of Mn in IrMn than that in (Pd)PtMn. During operation with a high sensing current, the temperature of the spin-valve element often goes up to 150 °C or more. Therefore, T_B of the IrMn-based spin valves, as shown in Fig. 4.42, is not sufficiently high for severe circumstances, as a further increase in the sensing current density is expected in future as mentioned in Sect. 4.3.1. The concentration of Mn in IrMn is much higher than that in PtMn. The diffusion of Mn from IrMn to other layers causes a deterioration of the MR properties. This problem is improved by using a SAF structure as the Ru plays the role of a barrier against the diffusion of Mn [57], though the SAF structure could not raise the T_B.

Spin valves using oxide antiferromagnets such as NiO [7] and α-Fe_2O_3 [8] show large MR ratios owing to specular scattering as mentioned in Sect. 4.2.2. For example, NiO(50)/α-Fe_2O_3(2)/NiFe(1.5)/CoFe(1.5)/Cu(2.5)/CoFe(2)/NiFe(4) spin valve [58] shows an MR ratio of 12% and a ΔRs of 2 Ω/□ with an exchange bias field, H_{ex}, of 400 Oe. However the T_B is about 220 °C and is not sufficiently high for practical application to MR heads. The critical (or minimum) thickness of NiO as the pinning layer is 40~50 nm, which is too thick for the spin valves used for MR heads with shield gaps narrower than 100 nm, at present. The α-Fe_2O_3 based-spin valves also show very large MR ratios such as 18~28% as mentioned in Sect. 4.2.2, though the thickness of α-Fe_2O_3 is 50 nm. The thickness of α-Fe_2O_3 can be reduced from 50 nm to 10 nm [59], but this causes a decrease of H_{ex} The SAF structured spin valves having α-Fe_2O_3 as the pinning layer [60] solves this problem. Figure 4.43 shows an MR curve for the SAF structured α-Fe_2O_3(20)/Co(2)/Ru(0.7)/Co(2)/Cu(2)/Co(1)/NiFe(5)/Ta(3) spin valve. The film shows an MR ratio of 10% and a ΔRs of 2.3 Ω/ with a sufficiently large H_{ex}. The free layer shows an excellent soft magnetic property with a coercivity of 0.5 Oe This may originate in the small surface roughness of the 20 nm thick α-Fe_2O_3 [59], which is effective for reducing the orange peel coupling between the free layer and the pinned layer. The insertion of the Ru layer causes a reduction of MFP of the

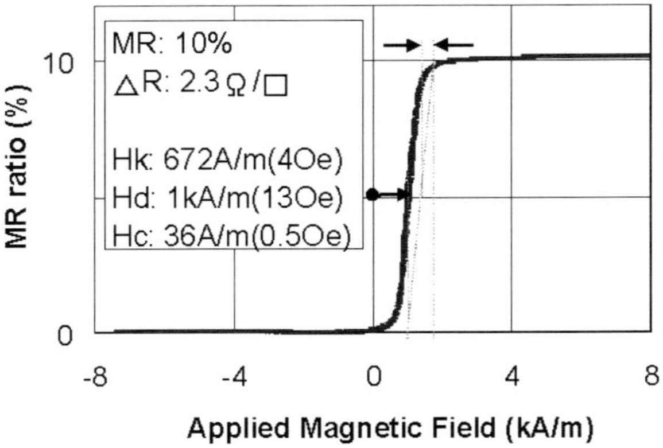

Fig. 4.43. MR curve of the SAF structured α-Fe_2O_3/Co/Ru/Co/Cu/Co/NiFe/Ta spin valve

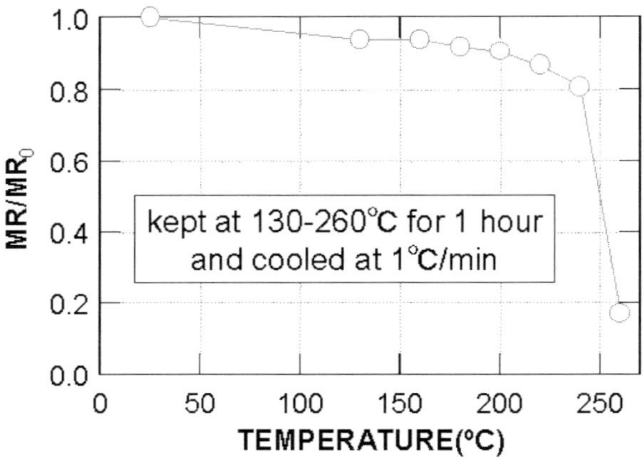

Fig. 4.44. Dependence of the MR ratios on the temperature of the rotational field annealing in α-Fe_2O_3(20)/CoFe(2)/Ru(0.7)/CoFe(2)/Cu(2)/CoFe(0.5)/NiFe(3)/Ta(3)

conduction electrons in the spin-valve, resulting in a decrease of the specular scattering effect at the α-Fe_2O_3/Co interface. This is the reason why the MR ratio of the SAF structured spin-valve is smaller than those of conventional ones. The total thickness of the metallic layers is only 15.7 nm. The α-Fe_2O_3 could be used as part of the insulator in the shield gap. Therefore, this spin valve is suitable for MR heads having ultra-narrow shield gaps. There remains one problem to be solved for this spin valve. Figure 4.44 shows the change of MR ratio of the spin valve after annealing with a rotational field of 100 Oe. The T_B of the spin-valve film is estimated at about 250 °C from the data, which is not sufficient for practical application to MR heads at present.

At present, spin-valve structured GMR films having Cu as the nonmagnetic layer (or separator) are used for read heads as mentioned above. But TMR films having Al_2O_3 as the nonmagnetic layer are another candidate for read heads for ultra- high-density recording, as TMR films show larger MR ratios and ΔRs, which enables larger output in the read heads. Details of TMR head will be discussed in Sect. 5.2.

References

[1] B. Dieny, V. S. Speriosu, S. S. P. Parkin, B. A. Gurney, D. R. Wilhoit and D. Mauri: *Phys. Rev. B* **43**, 1297 (1991).

[2] M Saito *et al*: *Digests of INTERMAG'97*, HA-06 (1997).

[3] C. Tang, N. Heiman and K. Lee: *J. Apply. phys.*, **52**, 2471 (1981).

[4] K. Hoshino, R. Nakatani, H. Hoshiya, Y. Sugita, and S. Tunahima: *Jpn. J. Appl. Phys.*, **35**, 607 (1996).

[5] T. Lin, C. Tang, R. E. Fontana and J. Howard: *IEEE Trans. on Mag.*, **31**, 2585 (1995).

[6] M. Saito, Y. Kakihara, T. Watanabe, and N. Hasegawa: *Digests of 20th Annual Conference on Magnetics in Japan*, 141 (1996).

[7] S. Soeya, S. Tadokoro, T. Imagawa, and M. Fukuyama: *J. Appl. Phys.*, **74**, 6297 (1993).

[8] W. C. Cain, W. H. Meikeljohn and H. Kryder: *J. Appl. Phys.*, **61**, 4170 (1987).

[9] H. N. Fuke, K. Sato, Y. Kamiguchi, H. Iwasaki, and M. Sahashi: *J. Appl Phys.*, **81**, 4004 (1997).

[10] W. F. Egelhoff, Jr., J. T .Ha, R. D. K. Misra, Y. Kadmon, J. Nir, C. J. Powell, M. D. Stiles, and R. D. McMichael; C.-L. Lin, J. M. Siverstsen, and J. H. Judy; T. Takano and A. E. Berkowitz; T. C. Anthony and J. A. Brug: *J. Appl. Phys.* **78**, 273 (1995).

[11] N. Hasegawa, A. Makino, F. Koike, and K. Ikarashi: *IEEE. Trans. Magn.*, **32**, 4618 (1996).

[12] W. F. Egelhoff, Jr., P. J. Chen, C. J. Powell, M. D. Stiles, R. D. McMichael, J. H. Judy, K. Takano and A. E. Berkowitz: *J. Appl. Phys.* **82**, 6142 (1997).

[13] Y. Sugita, Y. Kawawake, M. Satomi, and H. Sakakima: *Jpn. J Appl. Phys.* **37**, 5984 (1998).

[14] H. Sakakima, Y. Sugita, M. Satomi, and Y. Kawawake: *J. Mag. Mag. Mater.* **198-199**, 9 (1999).

[15] Y. Kawawake, Y. Sugita, M. Satomi, and H. Sakakima: *J. Appl. Phys.* **85**, 5024 (1999).

[16] J. L. Lee and M. K. Kryder: *J. Appl. Phys.* **83**, 3720 (1998).

[17] S. S. P. Parkin, N. More, and K. P. Roche: *Phys. Rev. Lett.*, **64**, 2304 (1990).

[18] G. Rupp, K. Schuster: *J. Magn. Magn. Mate..*, **121**, 416 (1993).

[19] H. C. Tong, X. Shi. F. Liu, C. Qian, Z. W. Dong, X. Yang, R. Barr. L. Miloslavsky, S. Zhou, J. Perlas, P. Prabhu, M. Mao, S. Funada, M. Gibbons, Q. Leng, J. G. Zhu, and S. Dey: *IEEE. Trans. Magn.* **35**, 2574 (1999).

[20] H. C. Tong, X. Shi. F. Liu, C. Qian, Z. W. Dong, X. Yang, R. Barr. L. Miloslavsky, S. Zhou, J. Perlas, P. Prabhu, M. Mao, S. Funada, M. Gibbons, J. G. Zhu, and S. Dey: *Digest of InterMag 99.*, AB-01 (1999).

[21] M. Saito, N. Hasegawa, K. Tanaka, Y. Ide, F. Koike, and T. Kuriyama: *Digest of 23rd Annual Conference on Magnetics in Japan.*, 417 (1999).

[22] A. V. Pohm, B. A. Everitt., R. S. Beech, J. M. Daugton: *IEEE Trans. Magn.* **33**, 3280 (1997).

[23] Y. Irie, H. Sakakima, M. Satomi, and Y. Kawawake: *Jpn. J. Appl. Phys.* **34**, L415 (1995).
[24] D. D. Tang, P. K. Wang, V. S. Sperosu, S. Le, and K. K. Kung: *IEEE Trans. Magn.* **31**, 3206 (1995).
[25] T. Shinjo and H. Yamamoto: *J. Phys. Soc. Jpn.*, **59**, 3061 (1990).
[26] Y. Kawawake, H. Sakakima Y. Irie, and M. Satomi: *J. Appl. Phys.* **79**, 6231 (1996).
[27] H. Sakakima, Y. Irie, Y. Kawawake, and M. Satomi: *J. Magn. Magn. Mater.* **156**, 405 (1996).
[28] Y. Kawawake and H. Sakakima: *J. Magn. Magn. Mater.* **149**, L255 (1995).
[29] H. Sakakima and M. Satomi: *Jpn. J. Appl. Phys.* **33**, L1668 (1994).
[30] H, Sakakima, M. Satomi, and H. Adachi: *Digest of the spring meeting of the Japan Institute of Metals*, 226 (2000).
[31] M. Jimbo, K. Komiyama, H. Matue, S. Tsunashima, and S. Uchiyama: *Jpn. J. Appl. Phys.* **34**, L112 (1995).
[32] H. Sakakima, M. Satomi, Y. Irie, and Y. Kawawake: *J. Magn. Magn. Mater.* **165**, 108 (1997).
[33] B. Dieny, P. Humbert, V. S. Speriosu, and B. A. Gurney: *Phys. Rev. B* **45**, 806 (1992).
[34] B. A. Gurney, V. S. Speriosu, J.-P. Nozieres, H. Lefakis, D. R. Wilhoit and O. U. Need: *Phys. Rev. Lett.* **71**, 4023 (1993).
[35] Y. Kamiguchi, H. Yuasa, H. Fukuzawa, K. Koui, H. Iwasaki, and M. Sahashi: *The Digest of INTERMAG 1999*, DB-01 (1999).
[36] J. Barnas, A. Fuss, R. E. Camley, P. Grunberg and W. Zinn: *Phys. Rev. B* **42**, 8110 (1990).
[37] M. Saito, N. Hasegawa, F. Koike, H. Seki and T. Kuriyama: *Digest of 23rd Annual Conference on Magnetics in Japan.*, 309 (1998).
[38] H. J. M. Swagten, G. J. Strijkers, P. J. H. Bloemen, M. M. H. Willekens and W. J. M.de Jonge: *Phys. Rev. B* **53**, 9108 (1996).
[39] W. F. Egelhoff, Jr., P. J. Chen, C. J. Powell, M. D. Stiles, and R. D. McMichael; J. H. Judy; K. Takano, and A. E. Berkowitz; and J. M. Daughton: *IEEE Trans. Magn.* **33**, 3580 (1997).
[40] Y. Kawawake and H. Sakakima: *IEEE Trans. Magn.* **33**, 3538 (1997).
[41] Y. Kamiguchi *et al*: *Digest of INTERMAG'99*, DB-01 (1999).
[42] H. Sakakima, M. Satomi, Y. Sugita, and Y. Kawawake: *J. Magn. Magn. Mater.* **210**, L20 (2000).
[43] H. Sakakima, M. Satomi, E. Hirota and H. Adachi: *IEEE Trans. Magn.* **35**, 2958 (1999).
[44] H. Yuasa, Y. Kamiguchi, and M. Sahashi: *Digest of 23rd Annual Conference on Magnetics in Japan.*, 416 (1999).
[45] Y. Shimizu, L. Varga, S. Eguchi, and A. Tanaka: *Digest of InterMag'2000*, FA-07 (2000).
[46] V. S. Speriosu, B. A. Gurney, D. R. Wilhoit, and L. B. Brown: *Digest of InterMag'96*, AA-04 (1996).
[47] M. Saito, N. Hasegawa, K. Tanaka, Y. Ide, F. Koike, and T. Kuriyama: *Digest of 23rd Annual Conference on Magnetics in Japan.*, 187 (1999).
[48] M. A. M. Gijs, S. K. J. Lenczowski, J. B. Giesberg: *Phys. Rev. Lett.* **70**, 3343 (1993).
[49] M. Saito, N. Hasegawa, F. Koike, H. Saki and T. Kuriyama: *J. Appl. Phys.* **85**, 4928 (1999).

[50] H. Sakakima, M. Satomi, Y. Sugita, and Y. Kawawake: *Digest of InterMag' 2000*, FA-08 (2000).

[51] Y. Sugita, Y. Kawawake, M. Satomi, H. Sakakima, S. Mitani, and A. Aono: *Digest of 24th Annual Conference on Magnetics in Japan.*, 421 (2000).

[52] A. Tanaka, Y. Shimizu, H. Kishi, K. Nagasaka, C. Hou, and H. Fujiwara: *Materials Research Society Symp. Proc.* **517**, 25 (1998).

[53] M. Komuro, H. Hosoya, Y. Kawato, K. Nakamoto, K. Meguro, T. Okada, Y. Hamakawa, M. Fujiwara, H. Fukui, and Y. Sugita: *J. Magnetic. Soc. Jpn.* **22**, 509 (1998).

[54] H. Kanai, J. Kane, and T. Yamamoto: *J. Magnetic. Soc. Jpn.* **20**, 361 (1996).

[55] J. Hong, K. Aoshima, J. Kane, K. Noma, and H. Kanai: *Digest of InterMag'2000*, BQ-12 (2000).

[56] E. N. Abarra, M. Suzuki, and I. Okamoto: *Digest of InterMag'2000*, AA-01 (2000).

[57] G. W. Anderson, M. Pakala, and Y. Huai: *Digest of InterMag'2000, BQ-02* (2000).

[58] J. Fujita, K. Hayashi, M. Saito and M. Nakada: *Digest of 21st Annual Conference on Magnetics in Japan.*, 47 (1997).

[59] Y. Kawawake, Y. Sugita, M. Satomi, and H. Sakakima: *J. Appl. Phys.* **85**, 5024 (1999).

[60] Y. Sugita, Y. Kawawake, M. Satomi, and H. Sakakima: *IEEE Trans.Magn.* **35**, 2961 (1999).

5 Tunnel-Type GMR (TMR) Devices

5.1 Spin-Dependent Tunneling

5.1.1 Tunneling Magnetoresistance

Tunneling magnetoresistance (TMR) is observed for ferromagnetic spin tunneling junctions (MTJ) consisting of ferromagnetic–insulator–ferromagnetic layers[1–3]. When the insulating layer, usually referred to as the barrier layer, is very thin (the order of 1 nm), electrons can tunnel through this forbidden region as a result of the wave-like nature of electrons for a voltage applied between the two electrodes, and can only be described in terms of quantum mechanics. The basic principle of TMR is the dependence of the tunneling probability on the relative orientation of magnetization in the two ferromagnetic electrodes. The tunneling conductance is spin dependent due to the spin dependent density of states (DOS) at the Fermi level for ferromagnets. When the applied voltage is small enough for electrons near the Fermi level to tunnel, the tunneling conductance G in the MTJ can be written as (5.1) – (5.3) by neglecting the spin dependency of the tunneling probability [4],

$$G = R^{-1} = \Sigma_\sigma | T |^2 D_{1\sigma}(E_F) D_{2\sigma}(E_F), \tag{5.1}$$

$$| T |^2 \exp(-2 s\chi), \tag{5.2}$$

$$\chi = [8\pi^2 m^* (\phi - E_F) / h^2], \tag{5.3}$$

where R, T, $D_{i\sigma}(E_F)$ are the junction resistance, tunneling probability, DOS at the Fermi level for spin σ band of the i-th ferromagnet, respectively, and ϕ, s, χ and h are the barrier height and thickness, electron wave vector in the barrier and the Planck constant, respectively. The TMR is defined as

$$\mathrm{TMR} = (R_{AP} - R_P) / R_P, \tag{5.4}$$

where R_P and R_{AP} are the resistance for parallel and antiparallel spin configurations of two ferromagnetic electrodes, respectively. If we assume that spin is conserved during tunneling as shown in Fig.5.1, (5.1) – (5.4) lead to Julliere's model [1],

$$\mathrm{TMR} = 2 P_1 P_2 / (1 - P_1 P_2). \tag{5.5}$$

P_i is the spin polarization of tunneling electrons of ferromagnet i and is given by

$$P_i = (D_{i\uparrow} - D_{i\downarrow}) / (D_{i\uparrow} + D_{i\downarrow}). \tag{5.6}$$

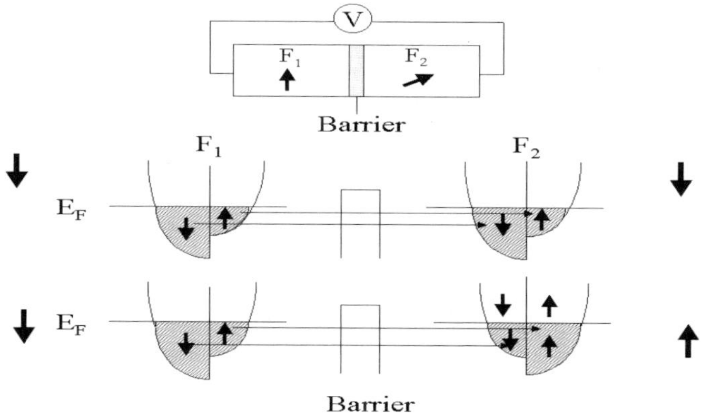

Fig. 5.1 A schematic model for spin–dependent tunneling

TMR can vary over a wide range depending on the P values of ferromagnets as verified by (5.5) and (5.6). For vanishing polarization of one of the electrodes the magnetoresistance disappears, while for full polarization of the tunneling electrons of both electrodes (P_1=P_2=1) the effect will become infinitely large corresponding to a value of 1 in (5.5). Unfortunately, determining the spin polarization at the Fermi energy of a ferromagnet is not easy. A typical transition metal ferromagnet has two components to its electronic structure; narrow d-bands that may be fully or partially spin polarized and broad s-bands with a lesser degree of spin polarization due to hybridization with the d-bands. The value of P is controlled by the extent to which these s- and d-bands cross the Fermi energy. If the orbital character at the Fermi surface of the ferromagnetic metal is primarily d-like, then the spin polarization will be high. If the orbital character, however, is s-like or s-d hybridized, then the spin polarization can be low or high depending on the detail of the electronic structure. The magnetization of a material may show that all the electronic spins associated with the d orbitals are aligned but the spin polarization at E_F can be depressed. On the other hand, metallic oxide ferromgnets, for example, have a greater opportunity for high spin polarization because of the predominance of the d orbital character at E_F.

Measuring the spin polarization requires a spectroscopic technique that can discriminate between the spin-up and spin-down electrons near E_F. Spin polarized photoemission spectroscopy is technically capable of providing the most direct measurement of P, but lacks the necessary energy resolution (~1 meV). An effective alternative to photoemission is the use of spin polarized tunneling in a planar junction geometry which does allow the electronic spectrum near E_F to be probed with sub-meV energy resolution.

Tedrow and Meservey [5] pioneered this technique by making superconductor/insulator/ferromagnet junctions and Zeeman splitting the superconductor's strongly peaked single–particle excitation spectrum by the application of a magnetic field. The spin-splitting of the quasiparticle density of states in a superconductor by the application of a magnetic field allows probing the spin polarization of tunneling electrons from the ferromagnet, which is schematically shown in Fig. 5.2. The resulting spectrum of the superconductor roughly corresponds to two fully spin polarized peaks (neglecting spin–orbit coupling effects) that can be used to detect the spin polarization of a current from the ferromagnetic film. Another method of measuring spin polarization of a metal was developed recently [6], which is a metallic point contact between the sample and the superconductor using Andreev reflection. Unlike a tunnel junction, a metallic contact allows coherent two-particle transfer at the interface between the normal metal and the superconductor. The electronic transport properties at the point contact measures the conversion between superconducting pairs and the single particle charge carriers of the metal. The conversion of normal current to supercurrent at a metallic interface is called Andreev reflection.

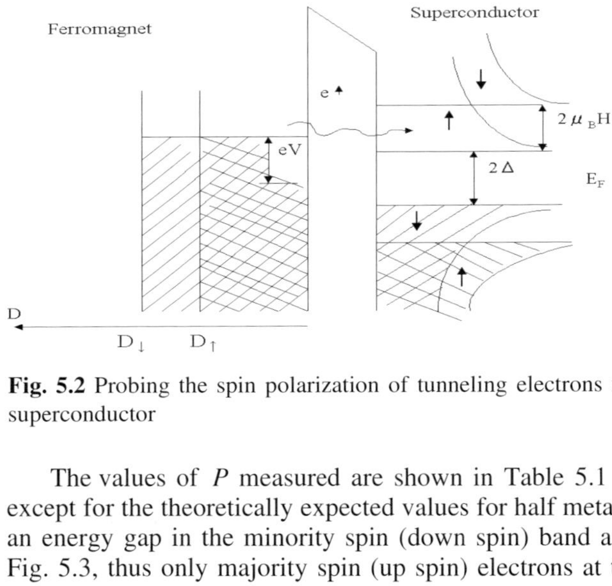

Fig. 5.2 Probing the spin polarization of tunneling electrons from the ferromagnet to the superconductor

The values of P measured are shown in Table 5.1 for various ferromagnets except for the theoretically expected values for half metals with $P = 1$, which have an energy gap in the minority spin (down spin) band as shown schematically in Fig. 5.3, thus only majority spin (up spin) electrons at the Fermi level. The spin polarization of tunneling electrons seems to be nearly proportional to the magnetic moment μ of the electrode as shown for Ni-Fe alloys in Fig. 5.4 [5], while it is not always proportional to μ of the ferromagnetic electrode as shown in Fig. 5.5 [7], which exhibits TMR and μ^2 as a function of the composition of Fe-Co alloy electrodes used for the junctions.

Table 5.1 Spin polarization of various magnetic materials

Magnetic materials	Spin polarization	Magnetic materials	Spin polarization
Fe	+0.44	NiMnSb	1, +0.58
Co	+0.35	PtMnSb	1
Ni	+0.23	CrO_2	1
$Ni_{80}Fe_{20}$	+0.25, +0.45	Fe_3O_4	1
$Fe_{50}Co_{50}$	+0.53	(La-Sr)MnO3	1

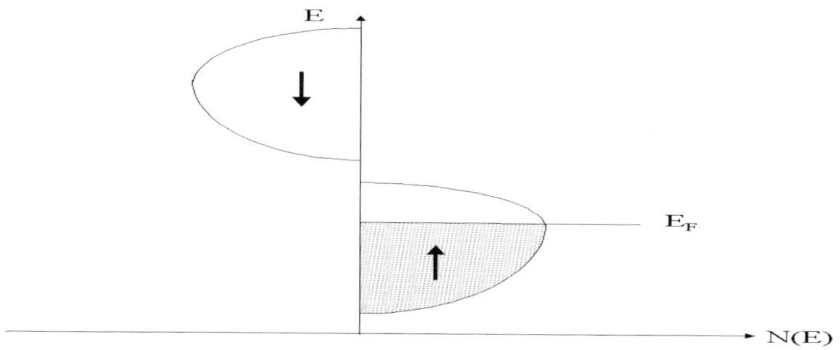

Fig. 5.3 Schematic energy band structure for a half metal

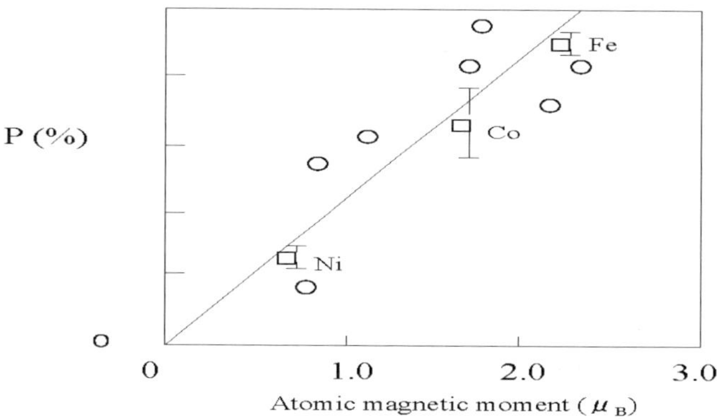

Fig. 5.4 Spin polarization of tunneling electrons vs the magnetic moment μ of the electrode

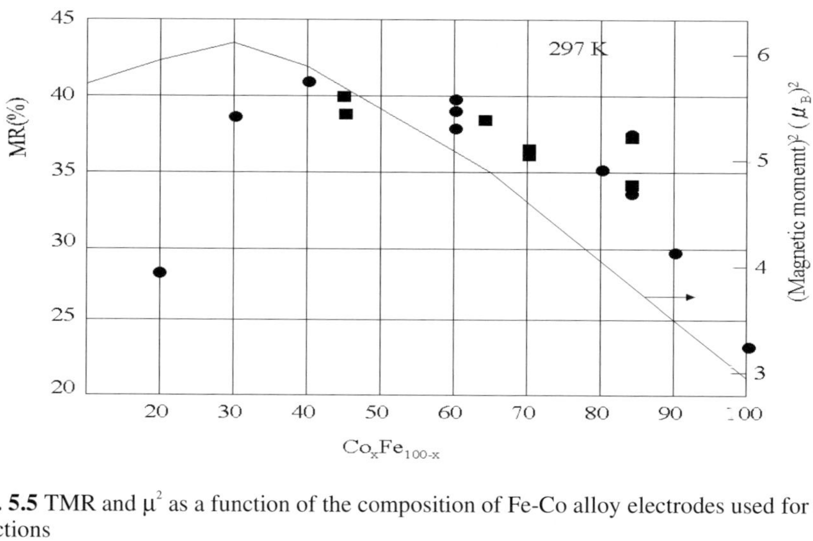

Fig. 5.5 TMR and μ^2 as a function of the composition of Fe-Co alloy electrodes used for the junctions

Equation (5.5) assumes that all electrons at the Fermi level have the same probability of tunneling and also neglects the influence of the barrier. In fact, however, TMR depends on tunneling barrier height as shown in Fig. 5.6 for the MTJ using CoFe electrodes. The TMR is larger for the higher barrier height. The barrier height can be estimated by Simmons' expressions (5.7) – (5.10) [8]

$$J = \beta\,(V + V^3), \tag{5.7}$$

$$\beta = [3e^2(2m\phi)^{1/2}/2h^2s]\exp(-A\phi^{1/2})\ [\Omega^{-1}cm^2], \tag{5.8}$$

$$\gamma = (Ae)^2/96\phi - Ae^2/32\ \phi^{1/4}\ , \tag{5.9}$$

$$A = 4\pi(2m)^{1/2}\ s/h = 1.027[\text{Å}/(eV)^{1/2}], \tag{5.10}$$

for lower bias voltage, where J,ϕ and s are current density [A/cm^2], barrier height [eV] and barrier thickness [cm], respectively. For junctions with large tunneling barrier height the model well explains the experimental results of TMR as shown in Fig. 5.7, which exhibits experimental TMR as a function of $2P_1P_2/(1-P_1P_2)$ for various junctions with different magnetic electrodes, where the experimentally obtained P was used. However, there is a serious failure in Julliere's model, that is, the sign of the spin polarization is different from experiments. The tunneling experiments of Tedrow and Meservey exhibited positive spin polarization for all the analyzed transition metals [9], in contrast with experiments. In transition metals like Co or Ni a negative spin polarization is expected because their 3d–band is negatively polarized at the Fermi level. To explain these results it has been claimed for instance that only some itinerant d electrons tunnel or only the s elect-

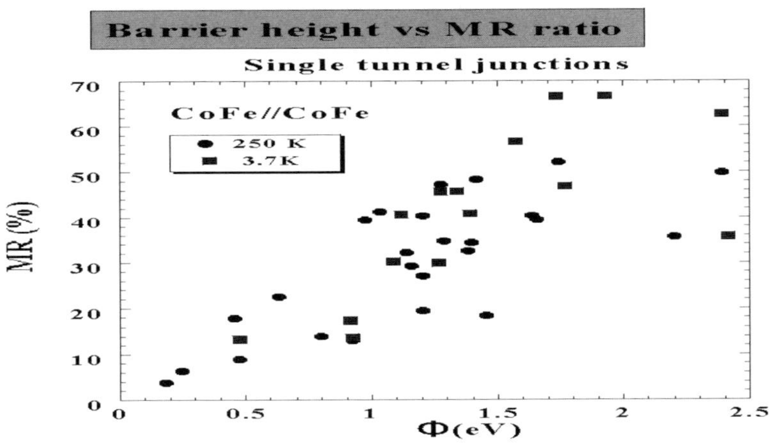

Fig. 5. 6 TMR as a function of barrier height at 250K and 3.7 K

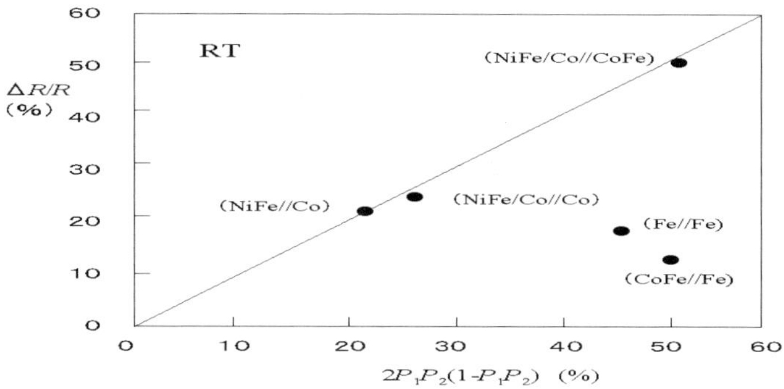

Fig. 5.7 Experimental TMR as a function of $2P_1P_2/(1-P_1P_2)$ for various junctions with different magnetic electrodes, where the experimentally obtained P was used

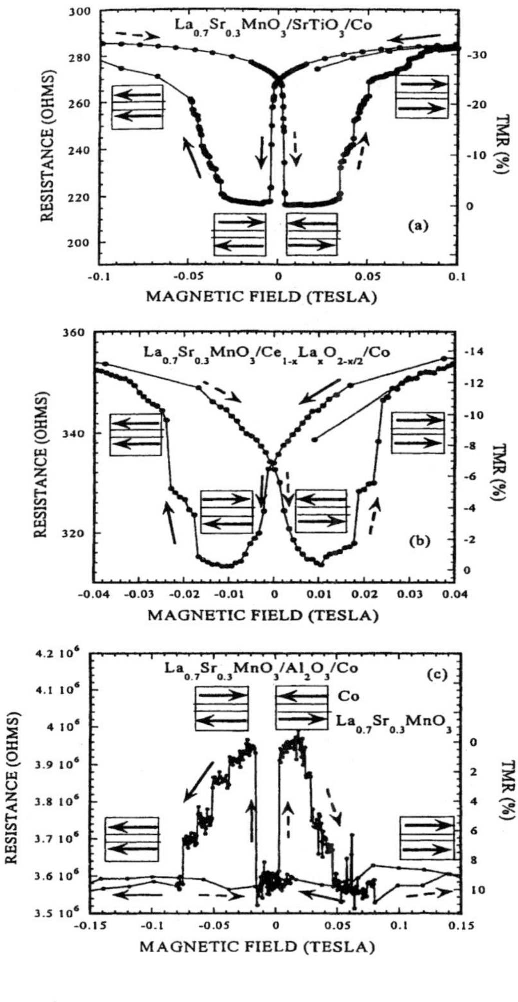

Fig. 5.8 TMR vs magnetic field for junctions consisting of ferromagnetic transition metal/insulator/half metal. The sign of the spin polarization depends on the barrier materials

rons, with lower DOS but positive spin polarization, can tunnel. All these explanations are unsatisfactory and the results of Tedrow and Meservey have remained a mystery for more than 20 years. Recently, Teresa *et al.* [10], however, demonstrated in experiments for junctions consisting of ferromagnetic transition metal/insulator/half metal that the sign of the spin polarization depends on the barrier materials as shown in Fig. 5.8, and that selection by the barrier of the character of the tunneling electrons can be explained by the character of the bonding between the electrode and the barrier.

5.1.2 Fabrication of Spin Tunneling Junctions

Spin tunneling junctions are usually fabricated by sputtering using metal masks to form cross–pattern junctions as shown in Fig. 5.9 using an Al_2O_3 barrier. An Al_2O_3 barrier layer is deposited in a two-step process. In the first step a thin metallic Al film is deposited. A metal is deposited because metal-on-metal wetting is generally much better than insulating-on-metal wetting. A smooth, homogeneous, and completely covering Al layer is obtained by sputter-deposition. In the second step the metallic Al layer is oxidized, for example, in a reactive oxygen plasma [3] or a natural oxygen atmosphere to form Al_2O_3 [2, 11].

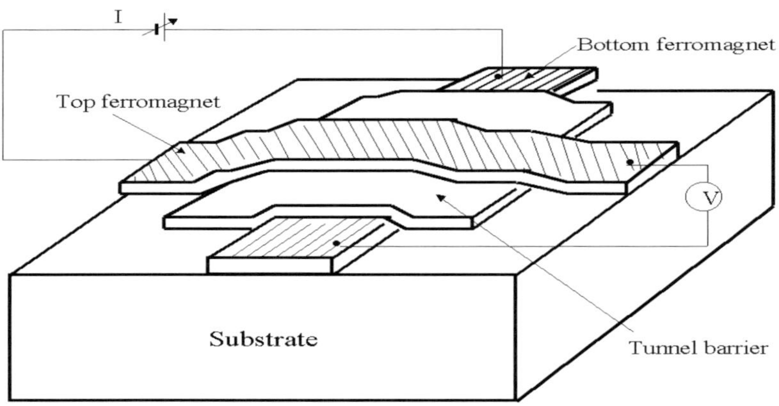

Fig. 5.9 Schematic cross–pattern junction fabricated by using metal masks

Most work on the MTJ has involved structures with two ferromagnetic layers composed of ferromagnetic materials with different coercive forces. At intermediate magnetic fields the magnetic moments of the two layers can be arranged to be approximately antiparallel to one another. However, it is advantageous to exchange bias one of the ferromagnetic layers with an antiferromagnetic layer like the spin-valve GMR structure which thereby leads to improved control of the magnetic alignment of the magnetic layers. A typical resistance vs magnetic field curve in a low applied field range at RT is shown in Fig. 5.10 for the as-deposited $Ni_{80}Fe_{20}$ (5 nm)/$Ir_{23}Mn_{77}$(12 nm)/CoFe (3 nm)/Al_2O_3/CoFe(5 nm)/$Ni_{80}Fe_{20}$(15 nm) spin–valve–type junction using an IrMn antiferromagnet, fabricated on a thermally oxidized Si substrate. The $Ni_{80}Fe_{20}$ (hereafter, NiFe) underlayer was used to grow an $Ir_{23}Mn_{77}$ (hereafter, IrMn) layer with good crystalline quality. The magnetic double layers with different film thickness made of NiFe and CoFe are used for two ferromagnetic layers with different coercive forces and enhanced TMR. The curve exhibits a square hysteresis and a large TMR of 36% at RT, due to the perfect antiferromagnetic alignment between two ferromagnetic electrode spins.

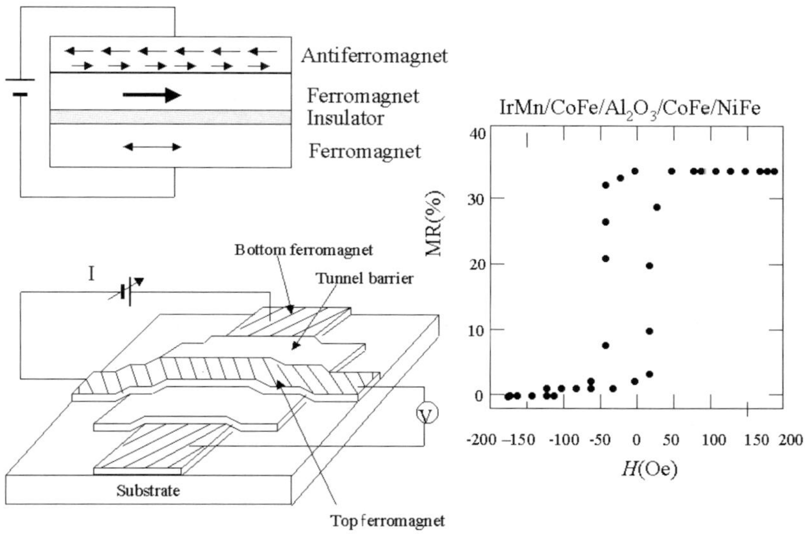

Fig. 5.10 Spin–valve–type junction structure and TMR curve for as-deposited $Ni_{80}Fe_{20}$ (5 nm)/$Ir_{23}Mn_{77}$ (12 nm)/CoFe (3 nm)/Al_2O_3/CoFe(5 nm)/$Ni_{80}Fe_{20}$(15 nm) using an IrMn antiferromagnet

The junction resistance depends on the Al_2O_3 barrier thickness as predicted by Eqs.(5.1) and (5.2). Figure 5.11 exhibits junction resistance vs Al_2O_3 layer thickness for Fe/Al_2O_3/CoFe junctions [12], in which the Al_2O_3 layer with a wedge shape was prepared by reactive evaporation of Al. The resistance is exponentially proportional to the Al_2O_3 layer as expected in (5.2).

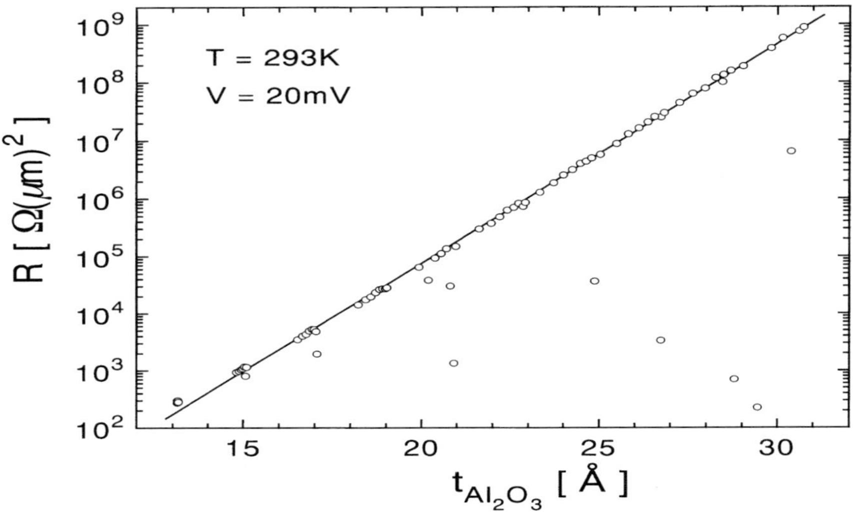

Fig. 5.11 Junction resistance as a function of Al_2O_3 thickness

5.1.3 Microfabrication of Spin Tunneling Junctions

A self-aligned lithographic process has been applied to make MTJ devices with dimensions on the micron–to–submicron scale using optical and electron-beam lithographic processes, respectively [13]. The patterning sequence is illustrated in Fig. 5.12. The fabrication process begins with the growth of a multi-layered film deposited at room temperature [Fig. 5.12(a)]. Next, the base electrode shapes are defined in resist and multilayer stack patterned by ion milling [Fig. 5.12(b)]. The junction areas are then defined by patterning the top electrode shape in resist and doing a second ion milling. This milling is timed to stop at the top surface of the base electrode [Fig. 5.12(c)]. Next, the junctions are coated with SiO_2 by rf magnetron sputtering [Fig. 5.12(d)]. This deposition is carried out in two steps to produce a virtually pin-hole free insulation layer. The resist used for the previous ion milling step is left on the wafer during the SiO_2 deposition so that it could be used as a lift-off stencil for self-aligned contact holes to the top junction electrodes [Fig.5.12 (e)]. Finally, a metallization layer of Au is deposited and patterned by lift-off to make connections to the top junction electrodes [Fig. 5.12 (f)].

In Fig. 5.13 TMR and junction resistance R are shown as a function of nominal junction area in a log-log plot for devices with Co-free layers patterned with both e-beam and optical lithographies [13]. The TMR is essentially independent of the junction areas although there are scatter of R and TMR in the figure. The straight line in Fig. 5.13 represents linear scaling between R and the area as expected from electron tunneling. The data generally follow the linear scaling relation.

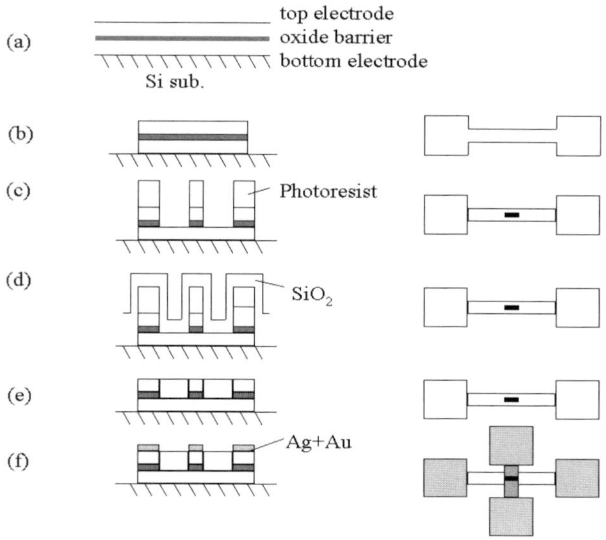

Fig. 5.12 A self aligned lithographic process applied to make MTJ devices

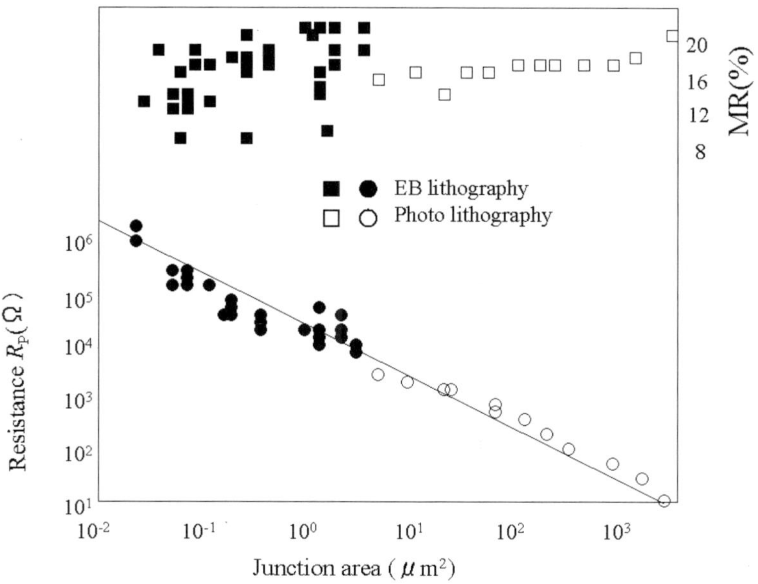

Fig. 5.13 Junction resistance and TMR as a function of junction area

5.1.4 Annealing Properties

A very important parameter for applications of MTJ is the thermal stability of the junctions. Annealing at an elevated temperature in a magnetic field in vacuum is effective for improving TMR. Figure 5.14 exhibits annealing temperature dependence of the TMR and junction resistance for the NiFe/IrMn/CoFe/Al_2O_3/CoFe/NiFe junction, which was prepared using ion beam sputtering [14]. The TMR increases at first with increasing annealing temperature and takes a maximum value of 39 ~ 41% at 290 °C while the resistance–area product $R\times A$ decreases. Further anneals at higher temperatures show a steep decrease of TMR. The improvement of the TMR at lower temperatures has been attributed to that an unoxidized Al interface layer is oxidized with the oxygen atoms from the oxidized Al and oxidized Co by annealing [15]. The decrease of TMR due to annealing at higher temperatures has been explained as diffusion of Mn from the antiferromagnetic layer through thin exchange-biased ferromagnetic layers to the tunnel barrier [16]. In fact, for an MTJ using Co/Ru/CoFe artificial ferromagnet systems as hard magnetic electrode, exchange-biased with IrMn antiferromagnet thermal stability is enhanced up to 350 °C at least [17].

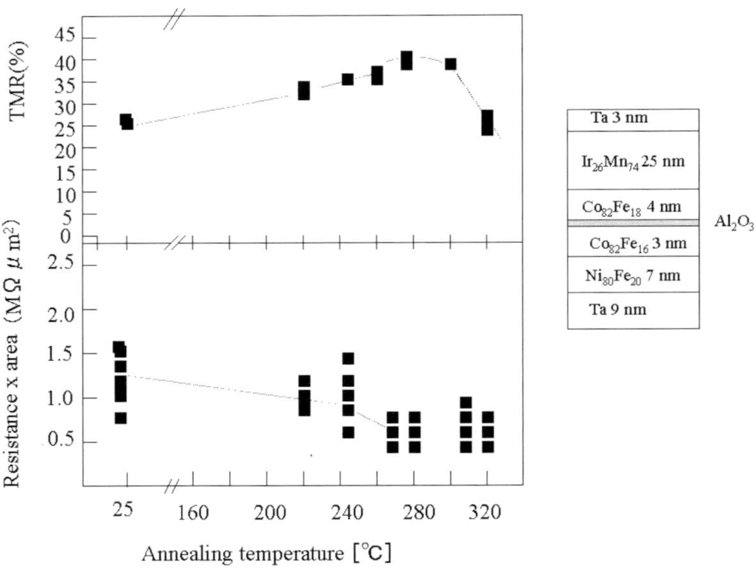

Fig. 5.14 Annealing temperature dependence of the TMR and junction resistance for the NiFe/IrMn/CoFe/Al2O3/CoFe/NiFe junction

5.1.5 Temperature Dependence of TMR

The observed TMR has reached nearly the optimum values expected from Julliere's model at low temperature. However, even with the best junctions there is a significant decrease in TMR at RT as compared to values at 4.2 or 77 K. Although the strength of the temperature dependence varies, it is present irrespective of the type of tunnel barrier and ferromagnetic electrodes. There are some candidates with regards to temperature dependence of TMR. First, it is elastic direct tunneling which varies slightly with temperature due to broadening of the Fermi distributions in the electrodes. This gives for the tunneling conductance [18], where G_0 is

$$G_T = G_0 \, C \, T \, / \, \sin (CT). \tag{5.11}$$

is a constant and C = 1.387 x 10^{-4} $s / \sqrt{\phi}$, with the barrier width (s) in $\AA$ and barrier height (ϕ) in eV. For typical barrier parameters, G_T at 300 K is only a few percent higher than at $T = 0$ K.

Second is a variation of electrode polarizations P with T. P in the case of alloys scales approximately with the magnetic moment of the alloy as its composition varies. A logical extension of this proportionality is to adopt a polarization that varies with T as does the magnetization. Magnetization versus T has been extensively studied, and is described fairly well by thermal excitation of spin waves for T below the Curie temperature. This produces a term proportional to the magnetization. For the polarization, thus, the following expression is given [18]

$$P(T) = P_0 (1 - \alpha T^{3/2}). \tag{5.12}$$

The material-dependent constant α is different for bulk or surface, and is generally larger for the latter due to surface exchange softness.

Equations (5.5) and (5.12) suggest that TMR may hardly change with temperature below RT for MTJ with large barrier height and high Curie temperature, because both the G_T and P (T) are weakly temperature dependent for such conditions. Nevertheless, TMR, in general, decreases with increasing temperature below RT. For this fact the contribution of spin independent indirect tunneling has been proposed, in addition to spin-dependent direct tunneling [19]. In this case, the tunneling conductance can be expressed as

$$G(\theta) = G_T \{ 1 + P_1 P_2 \cos(\theta) \} + G_{SI}. \tag{5.13}$$

This expression leads to

$$\Delta G = G(0) - G(\pi) = 2 \, G_T P_1 P_2, \tag{5.14}$$

$$\mathrm{TMR} = 2 \, G_T P_1 P_2 / [G_T (1 - P_1 P_2) + G_{SI}], \tag{5.15}$$

where θ is the angle between the magnetization directions of the two electrodes and G_{SI} is the spin-independent indirect tunneling conductance. Several physical mechanisms may be responsible for the G_{SI}. First of all, imperfections in the amorphous Al_2O_3 barrier may be present in sufficient density to provide a noticeable hopping conductance through the associated localized states. Other processes, for instance some ions of the electrodes may be misplaced in the Al_2O_3, producing states near the barrier interfaces. The above-mentioned mechanisms for G_{SI} do not rely on the ferromagnetic nature of the electrodes.

In Figs. 5.15 and 5.16 normalized TMR and ΔG are shown as a function of temperature, respectively for the IrMn/CoFe/Al_2O_3/CoFe/NiFe junctions with various barrier heights and thicknesses. The TMR decreases with increasing temperature, while ΔG is almost temperature independent below RT. The latter can be

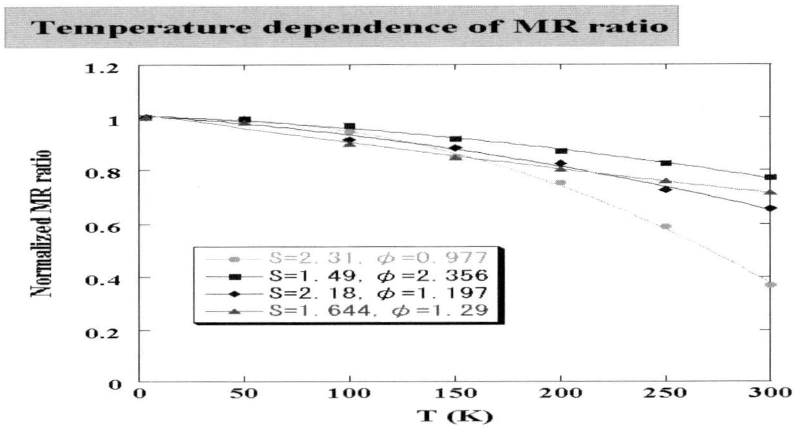

Fig. 5.15 Normalized TMR as a function of temperature for the IrMn/CoFe/Al_2O_3/CoFe/NiFe junctions with various barrier heights and thicknesses

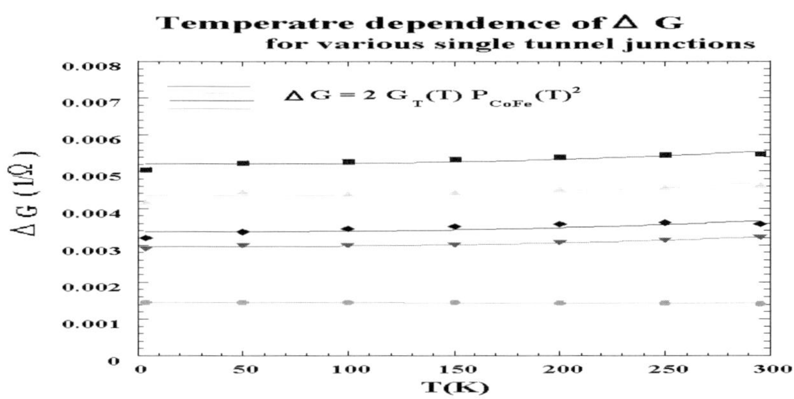

Fig. 5.16 Conductance change ΔG as a function of temperature for the IrMn/CoFe/Al_2O_3/CoFe/ NiFe junctions with various barrier heights andthicknesses

understood by considering an almost temperature independent P due to the high Curie temperature of CoFe alloy in (5.14). Thus, the temperature dependence of TMR in Fig. 5.15 may be understood by the temperature dependence of G_{SI}. This suggests that the quality of the Al_2O_3 tunneling barrier is important to improve the temperature dependence of TMR, because the defect states in the barrier are uniformly distributed, both spatially and energetically, and the energetic distribution of the available defect states is governed by a Fermi-Dirac function [19] where E is the energy level of the defect state, E_c is the energy level of the barrier conduction band edge, k is the Boltzmann constant, and the T_{eff} is the effective barrier temperature. As suggested by (5.16), the density of available states increases exponentially as the energy level increases, which may explain the temperature dependence of the TMR.

$$f(E) = 1/[1+\exp(E_c-E)/kT_{eff}], \quad (5.16)$$

5.1.6 Bias Voltage Dependence of TMR

TMR, in general, largely decreases with increasing bias voltage [20]. The bias voltage dependence of TMR arises from the influence of the applied electric field on the barrier shape. Increasing the voltage increases the overall conductance and hence decreases the TMR. The calculated TMR using the free electron model falls much slower than experiment, especially at low bias voltage [21]. The energy dependence of the spin polarization due to band structure effects may also reduce the TMR at high voltage.

The part of the experimental large decrease in TMR can be attributed to magnon excitation in the magnetic layers, thereby randomizing the tunneling electron spins and increasing the total conductance [22]. This mechanism, however, can not always explain the experimental result that the bias voltage dependence of TMR depends on the barrier height and width. The bias voltage dependence of TMR at RT shown in Fig. 5.17 is for various junctions, prepared with different magnetic electrodes and different oxidization conditions of Al. The voltage V_h,

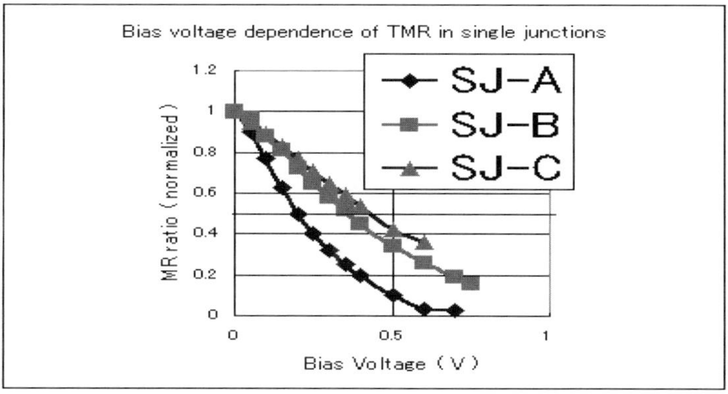

Fig. 5.17 Bias voltage dependence of TMR at RT for various junctions, prepared with different magnetic electrodes and different oxidization conditions of Al

where the TMR decreases to half of its zero bias value is 400 mV at most. It is sensitive to barrier height and barrier thickness, estimated by Simmon's expressions, as shown in Figs.5.18 and 5.19, respectively. V_h monotonically increases with increasing barrier height, while it decreases with increasing barrier thickness. This behavior may be understood by considering the indirect tunneling contribution through impurity states in the Al_2O_3 barrier, as in the case of the temperature dependence of TMR, because as suggested by (5.16), indirect tunneling, which is proportional to the defect state density, increases quickly at an increased bias voltage.

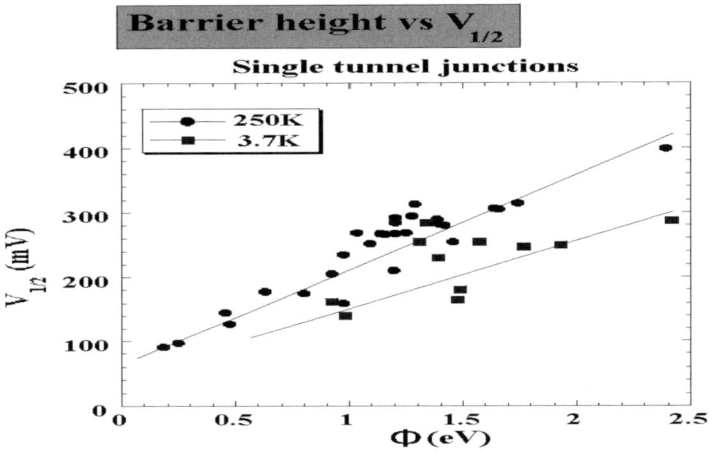

Fig. 5.18 V_h as a function of barrier height

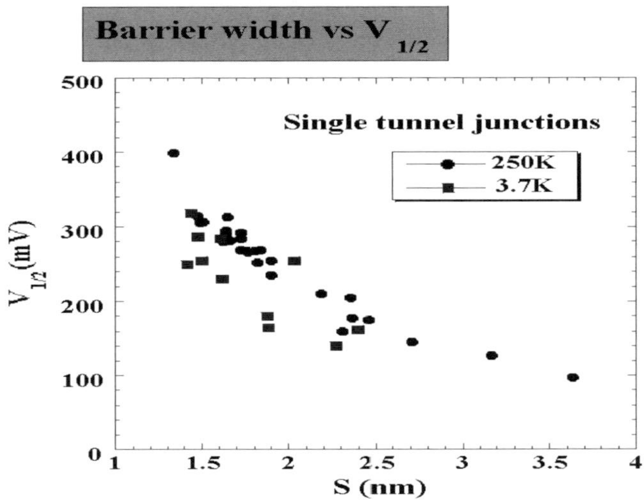

Fig. 5.19 V_h as a function of barrier thickness

5.1.7 Double Tunneling Junctions [23]–[25]

Spin–dependent double tunneling junctions have three magnetic layers and double barriers between two magnetic layers as shown schematically in Fig. 5.20, in which both sides of the magnetic layers are pinned by two exchange coupled antiferromagnetic layers, and the central soft magnetic layer spins can be switched by a low external field. If electrons tunnel sequentially through double barriers, TMR in double tunneling junctions is basically the same as that of single junctions. The maximum TMR value of over 40% has been attained at RT for the NiFe/IrMn/CoFe/Al_2O_3/Co_9Fe/Al_2O_3/CoFe/IrMn/NiFe double junctions [25], of which a cross sectional transmission electron microscopy image is shown in Fig. 5.21. The TMR value at RT may be larger than the value expected from Jurriere's model, taking account of the spin polarizations of the CoFe and Co_9Fe, and the temperature dependence of TMR. In fact the value is larger than that of the corresponding CoFe/Al_2O_3/Co_9Fe single junctions. The cause may be attributed to the concomitant tunneling, such as resonant tunneling, in addition to sequential tunneling in the double junctions, due to the thin magnetic Co_9Fe layers sandwiched between the two Al_2O_3 barriers.

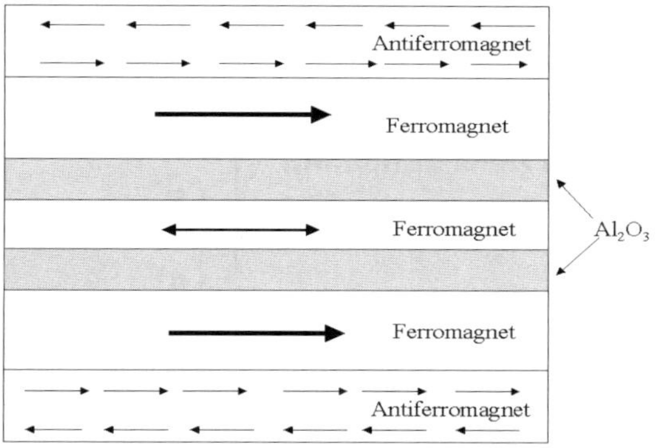

Fig. 5.20 A schematic double tunneling junction structure

Microfabrication of double tunneling junctions can be made without degrading the TMR using the same process as for single junctions. The TMR and the resistance of the double junctions fabricated by photolithography and ion milling are shown as a function of junction area in Fig. 5.22. The TMR is essentially constant for the junction area, while the resistance monotonically increases with decreasing junction area and decreases with decreasing Al_2O_3 barrier thickness. The optimum resistance of the double junction for devices such as the MRAM can be controlled by changing Al_2O_3 barrier thickness.

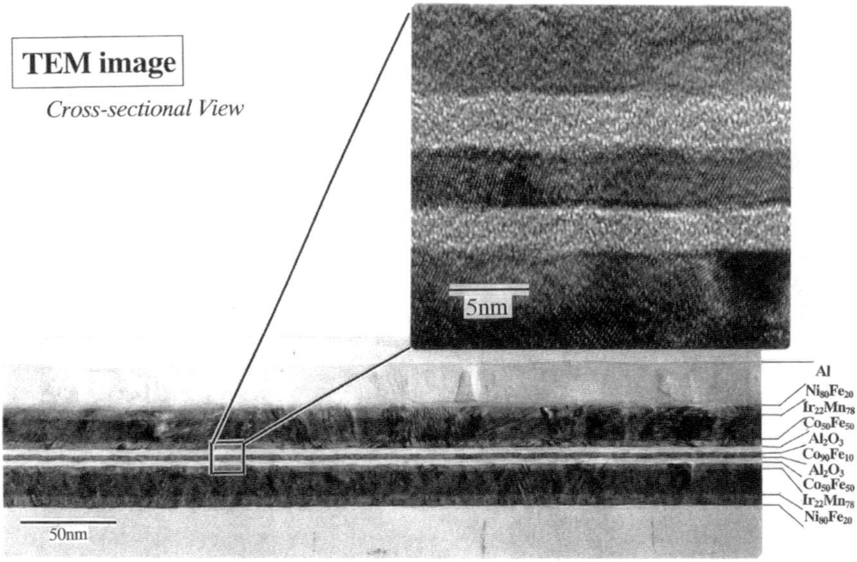

Fig. 5.21 Cross sectional transmission electron microscopy image for the double junction

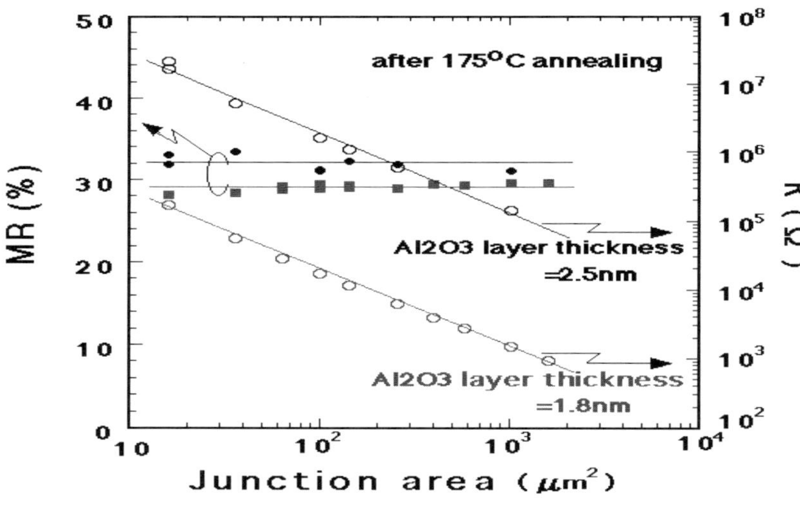

Fig. 5.22 The TMR and resistance of the double junctions fabricated by photolithography and ion milling as a function of junction area

It is expected that the bias voltage dependence of TMR in double junctions can be improved compared with that of single junctions, since the voltage drop across each barrier in a double junction is half of the applied voltage. Figure 5.23 exhibits the bias voltage dependence of TMR in double and single junctions, in which the TMR is normalized by zero bias value. The V_h is over 900 mV for the double junction, while it is 400 mV for the single junction. It is noted that the V_h in the double junction is over twice of the V_h of the single junction, which may also be attributed to the concomitant tunneling in the double junctions noted above.

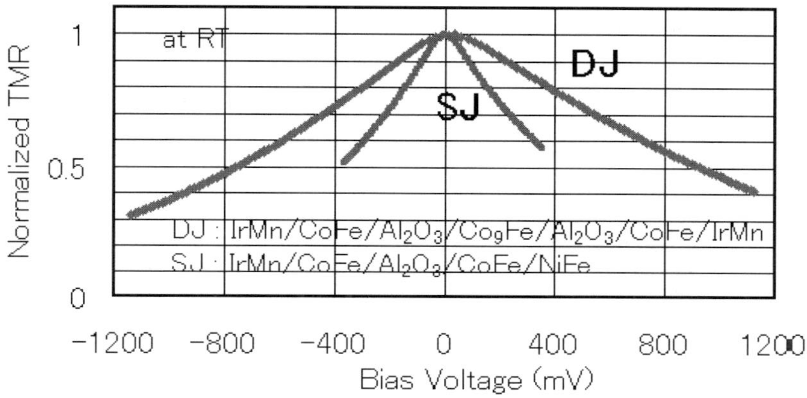

Fig. 5.23 Bias voltage dependence of TMR in the double junction compared with that of the single junction

5.2 Magnetic Read Heads of TMR

TMR films show larger ΔR and larger MR ratios than GMR films. Therefore, they have the potential for future read heads that enable high-density recording beyond 100 Gb/in^2. Recently, the development of TMR heads started in several companies [26–28]. Figure 5.24 shows a typical structure of the TMR head. The TMR element is composed of an antiferromagnetic layer, AF, a pinned layer, PL, an Al_2O_3 and a free layer, FL, and connected to the upper and lower shields through the nonmagnetic leads. Two hard magnetic layers, HM, for generating the bias field are placed on both side of the FL. The sense current flows perpendicular to the film plane of the TMR element. The upper and lower shields are used as part of the leads in this case. Therefore, it is not necessary to insulate the TMR element from the shields, which enables a narrower shield gap than in the case of GMR heads. Figure 5.25 shows the cross-sectional view of the TMR head that uses the free layer as a flux guide [26]. This structure is advantageous to avoid electrical shorting between the free layer and the pinned layer across the Al_2O_3 during the fabrication of the air-bearing surface of the TMR head. The TMR element used for the TMR head developed by TDK is composed of Ta(5)/NiFe(10)/Co(2)/AlO/CoFe/Ru/CoFe/PtMn (30)/Ta(5) [26] and the characteristics of the TMR head are summarized in Table 5.1. The output of the TMR head is much higher than that of the GMR head (~2.0 mV_{pp}), but the S/N ratios of the TMR and GMR heads are 35 dB and 32.2 dB, respectively. This result originates in the high resistance of the TMR element that increases the noise of the TMR head. Therefore, the reduction of the head resistance is also important to improve the S/N ratios of the TMR heads.

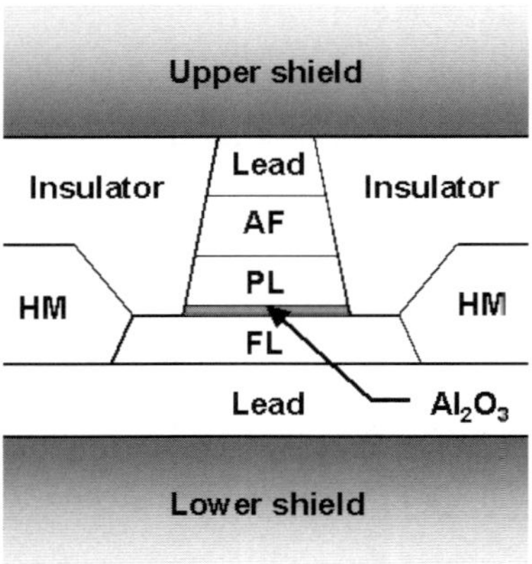

Fig. 5.24 Typical structure of the TMR head

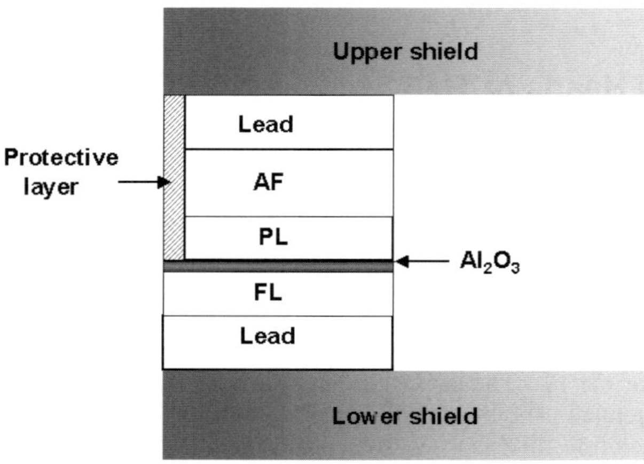

Fig. 5.25 Cross–sectional view of the TMR head

Table 5.1 Characteristics of a TMR head

MR ratio (at 300mV)	R	RA	T_W	Output
~20 %	106 Ω	30-35 $\Omega\mu m^2$	0.8 μm	5.74 mV_{pp}

The output of the read heads using TMR films as the MR element is given as

$$\begin{aligned} \Delta V &\propto I \bullet \Delta R \bullet \varphi/(M_F \bullet t_F) \\ &I \bullet \Delta R_A/(T_W \bullet T_H) \bullet \varphi/(M_F \bullet t_F) \\ &= J \bullet \Delta R_A \bullet \varphi/(M_F \bullet t_F) \end{aligned} \tag{5.17}$$

$$(J = I/(T_W \bullet T_H),\ \Delta R_A\text{: } \Omega\mu m^2)$$

where R_A is the resistance area product of the MTJ (magnetic tunneling junction). If we take the bias dependence of the TMR element into consideration, $I \bullet \Delta R$ in (5.17) becomes

$$I \bullet \Delta R = I \bullet R \bullet (\Delta R/R) = V_B \bullet \mathrm{MR}_0 \bullet (V_{max} - V_B)/V_{max}, \tag{5.18}$$

where MR_0 is the maximum MR ratio measured with almost zero bias voltage, V_B is the bias voltage and V_{max} is the voltage where the MR ratio becomes zero. Therefore, the output shows a maximum with $V_B=(1/2)V_{max} \equiv V_{1/2}$. Then the maximum output is given as

$$\Delta V \propto V_{1/2} \bullet \mathrm{MR}_0 \bullet \varphi/(M_F \bullet t_F), \tag{5.18$'$}$$

Therefore, the increase of $V_{1/2}$ is another important factor for the TMR element. Let us consider the changes in the output when the track pitch and the bit length are reduced to 1/5 and 1/2, respectively. Under the constant current condition, (5.17) becomes

$$\Delta V' \propto I \bullet \Delta R_A \bullet (25/2)/(M_F \bullet t_F), \tag{5.17$'$}$$

and under the constant current density condition, (5.17) becomes

$$\Delta V'' \propto J \bullet \Delta R_A \bullet (1/2)/(M_F \bullet t_F). \tag{5.17$''$}$$

We assume that T_H is also reduced to 1/5 like T_W. Comparing (5.17′) and (5.17″) for a TMR head with (4.15′) and (4.17′) for a GMR head, it may be easily understood that the TMR heads are more advantageous in the output than the GMR heads when the recording bit size becomes smaller. However, the resistance of the TMR head is drastically increased with the decrease of T_W, as the resistance is given as $\Delta R_A/(T_W \bullet T_H)$ and T_H is proportional to T_W. The resistance of the head-gimbals assembly is required to be smaller than 160 Ω to cope with high-frequency operation [26]. Therefore, it is desirable that the resistance of the TMR head be smaller than 100 Ω. If T_W and T_H become less than 0.1 mm, then R_A is required to be less than 1 $\Omega\mu m^2$. The Al_2O_3 tunneling barrier is formed by the oxidation of metallic Al. The smallest R_A at present is about 10 Ω obtained by the natural oxidation of 0.7 nm thick Al [27]. Therefore, it is required to form the Al_2O_3 barrier, prepared by natural oxidation of an ultra thin (0.5 nm) metallic Al layer, which can electrically and magnetically separate the free layer from the pinned layer in the TMR element. This is a big challenge for TMR heads.

References

[1] M. Julliere: *Phys. Lett.* **54A**, 225 (1975).
[2] T. Miyazaki and N. Tezuka: *J. Magn. Magn. Mater.* **139**, L231 (1995).
[3] J. S. Moodera, L. R. Kinder, T. M. Wong and R. Meservey: *Phys.Rev.Lett.* **74**, 3273 (1995).
[4] S. Maekawa and U. Gafvert: *IEEE Trans. Magn.* **MAG-18**, 707 (1982).
[5] P. M. Tedrow and R. Moservey: *Phys. Rev. Lett.* **26**, 192 (1971).
[6] R. J. Soulen Jr., J. M. Byers, M. S. Osofsky, B. Nadgorny, T. Ambrose, S. F. Cheng, P. R. Broussard, C. T. Tanaka, J. Nowak, J. S. Moodera, A. Barry and J. M. D. Coey: *Science*, **282**, 85 (1998).
[7] S. S. P. Parkin, K. P. Roche, M. G. Samant, P. M. Rice, R. B. Beyers, R. E. Scheuerline, E. J. O'Sullivan, S. L. Brown, J. Bucchigano, D. W. Abraham, Y. Lu, M. Rooks, P. L. Trouillou, R. A. Wanner and W. J. Gallargher: *J. Appl. Phys.* **85**, 5828 (1999).
[8] R. Moservey and P. M. Tedrow: Phys. Rep. 238, 173 (1994).
[9] J. G. Simmons: *J. Appl.Phys.* **35**, 2655 (1964).
[10] J. M. Teresa, A. Bathelemy, A. Fert, J. P. Cpnyour, R. Lyonner, F. Montaigne, P. Seneor and A. Vaures: *Phys. Rev. Lett.*, **82**, 4288 (1999).
[11] H. Tsuge and T. Mitsuzuka: *Appl. Phys. Lett.* **71**, 3296 (1997).
[12] Yuasa et al.: *Europhys. Lett.*, **52**, 344 (2000).
[13] W. J. Gallagher, S. S. P. Parkin, Y. Lu, X. P. Bian, A. Marley, K. P. Roche, R. A. Altman, S. Rishton, C. Jahnes, T. M. Shaw and G. Xiao: *J. Appl. Phys.***81**,3741 (1997).
[14] S. Cardoso, V. Gehanno, R. Ferreira and P. Freitas: *IEEE Trans. Magn*, **35**, 2952 (1999).
[15] M. Sato, H. Kikuchi and K. Kobayashi: *J. Appl. Phys.* **83**, 6691 (1998).
[16] M. G. Samant, J. Luning, J. Stohr and S. S. P. Parkin: *Appl. Phys. Lett.*, **76**, 3097 (2000).
[17] T. Dimopoulos C. Tiusan, K. Ounadjela, M. Hehn, H. A. M. van den Berg, Y. Henry and V. da Costa: *J. Appl. Phys.* **87**, 4685 (2000).
[18] C. He Shang, J. Nowak, R. Jansen and J. S. Moodera: *Phys. Rev.* **B58**, R2917 (1998).
[19] J. Zhang and R. M. White: *J. Appl. Phys.* **83**, 6512 (1998).
[20] J. S. Moodera, J. Nowak and R. J. M. van Veedonk: *Phys. Rev. Lett.*, **80**, 2941 (1998).
[21] A. M. Bratkovsky: *Phys. Rev.* **B56**, 2344 (1997).
[22] S. Zhang, P. Levy, A. C. Marley and S. S. P. Parkin: *Phys. Rev. Lett.*, **79**, 3744 (1997).
[23] K. Inomata, Y. Saito, K. Nakajima and M. Sagoi: *J. Appl. Phys.* **87**, 6064 (2000).
[24] Y. Saito, K. Nakajima, K. Tanaka and K. Inomata: *IEEE Trans. Magn.* **35**, 2904 (1999).
[25] Y. Saito, M. Amano, K. Nakajima, S. Takahashi, M. Sagoi and K. Inomata: *IEEE Trans Magn.* (2001) in press .
[26] K.Shimazawa, O.Redon, N.Kasahara, J.Sun, H.Morita, and M.Matsuzaki; *Digest of Intermag* 2000 FA-01 (2000).
[27] D.Song, J.Nowak, R.Larson, P.Kolbo, R.Chellew; *Digest of Intermag* 2000 FA-02 (2000).
[28] K.Hayashi, E.Fukami, K.Nagahara, M.Nakada, K.Ohashi; *Digest of Inter ag* 2000 FA-03 (2000).

6 Magnetic Random Access Memory (MRAM)

6.1 MRAM Devices and their Operation

6.1.1 Early MRAM Devices

Early magnetic random access memory used the natural hysteresis of magnetic materials to store data by using two or more sets of current•carrying wires or straps. Magnetic elements were arrayed so that only those which were to be written to received a combination of magnetic fields above a write threshold, while the other elements in the array did not change storage state. Most of today's MRAM concepts still use this technique.

Early ferrite core and magnetic thin film random access memories used inductive signals for determining the storage state ("1" or "0"). A magnetic field (current) was used to interrogate the memory element, and the polarity of the induced voltages in a sensing circuit depended on whether a "1" or "0" was stored. The first magnetoresistive readout scheme stored data in a magnetic body, which in turn produced a stray magnetic field that could be detected by a separate magnetoresistive sensing element. The concept was not high density because it was difficult to get a sufficiently large external stray field from a small magnetic storage cell.

The first MRAM devices utilizing magnetoresistance were developed in the mid–980s at Honeywell using the anisotropic magnetoresistance (AMR) effect, finding niche markets in satellite and military applications [1] due to the comparatively poor performance attributable to the low AMR of about 2%. AMR is a materials property whereby the angle which the current makes with respect to the magnetization in a ferromagnet affects the resistance of the sample. With recent advances in giant magnetoresistance (GMR) and TMR materials higher signal levels became available and renewed interest has arisen in MRAM fabrication. Now, MRAM possesses the attractive properties of nonvolatility, radiation hardness, nondestructive readout, low voltage, unlimited read and write endurance and high density. GMR and TMR materials exhibit a resistance change when an external magnetic field is applied and hysteresis in the magnetic switching curves, allowing their use in memory applications. In MRAM storing data is accomplished by applying magnetic fields and thereby causing a magnetic material in a cell to be magnetized into either of two possible memory states. Recalling data is accomplished by sensing resistance changes in the cell when the sensing current is supplied. The magnetic fields are created by passing currents through strip lines external to the magnetic structure (TMR type), or through the magnetic structures themselves (GMR type).

6.1.2 MRAM Devices Using GMR Cells

The discovery of GMR materials in 1988 gave hope for higher signals and faster access time. Magnetic films sandwiching a Cu layer and etched into stripes show an MR ratio of about 6%. Even with GMR materials, however, this cell had serious limitations. MRAM read access time was slower than for semiconductor memory because of the low MRAM sense signal. The invention of the pseudo-spin-valve (PSV) cell significantly improved signal levels, thus improving the read access time of MRAM while maintaining densities competitive with other solid state memory technologies. In the PSV cell not only was nearly all of the approximately 6% GMR available, but also the signal swing was plus or minus 6%, making the difference between a "0" and "1" about 12% of the cell resistance. This gave an eight times improvement over the original mode of operation and put MRAM on a much more even footing with semiconductor memory for read access time.

Figure 6.1 illustrates the conceptual MRAM memory cell structure using a PSV cell. The PSV cell providing GMR is made of two magnetic layers with different coercive forces, separated by a conductive nonmagnetic interlayer. This can be done by using two magnetic films of the same materials, but with different thicknesses. In that case, the thinner film switches at lower fields or the soft film and the thicker film switches at a higher field and is hard film. Thus, for the PSV at low field values the layer with the larger coercive force (hard layer) could be considered to be pinned while the layer with the smaller coercive force (soft layer) is reversible. In this structure the data are stored in the hard layer, while the switching of the soft layer can be used to modulate the resistance. Shape anisotropy provides the two-state energy well for storing "1"s and "0"s.

The PSV cell requires that the hard layer is switched for writing a bit. During the write operation, a forward word line current is applied. The resistance is lowest at the fields where the hard film switches to align with the soft film. The soft film acts as a means of reading the storage state, which is stored in the hard film. Without switching the hard film, the soft film can be manipulated to be parallel or antiparallel to the hard film. With a sequence of word fields which starts with a negative field and ends with a positive field, the resistance either rises or falls, depending on whether a "1" or "0" is stored. The direction of the sense line current determines which state the memory cell will be written to. In principle, this can be accomplished in a two-dimensional array with coincident currents which supply a magnetic field in the same direction, or with coincident currents which produce fields that are orthogonal. For PSV cells the sense current could also supply magnetic fields which aid switching.

The GMR MRAM architecture is a series of N resistors as shown in Fig. 6.2(a), while the TMR MRAM architecture consists of N parallel resistors as shown in Fig.6.2(b). The PSV sensing mode doubles the signal provided by the GMR devices but leads to higher write currents compared with MTJ devices, since the data

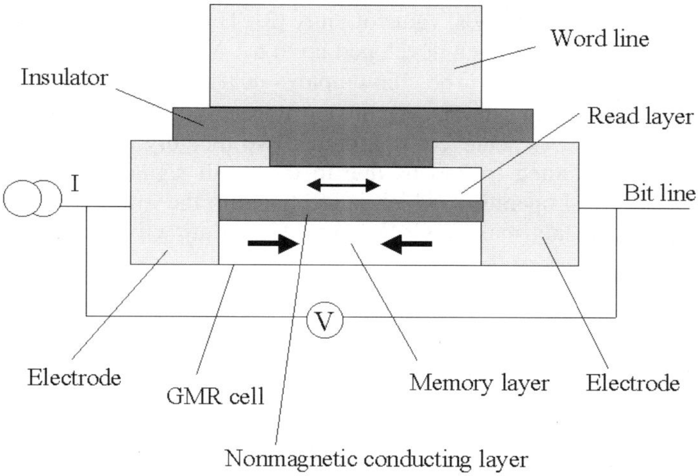

Fig. 6.1 Conceptual MRAM memory cell structure using a PSV cell

are stored in the harder of the two ferromagnetic layers. In the PSV case, the output signals is $I_s\Delta R$, but the noise increases as $(N)^{1/2}$, where I_s is the sensing current. Thus, the signal-to-noise ratio is given by

$$S/N = I_s R(GMR)/[1.26\text{x}10^{-6}(F)(NR)]^{1/2},$$

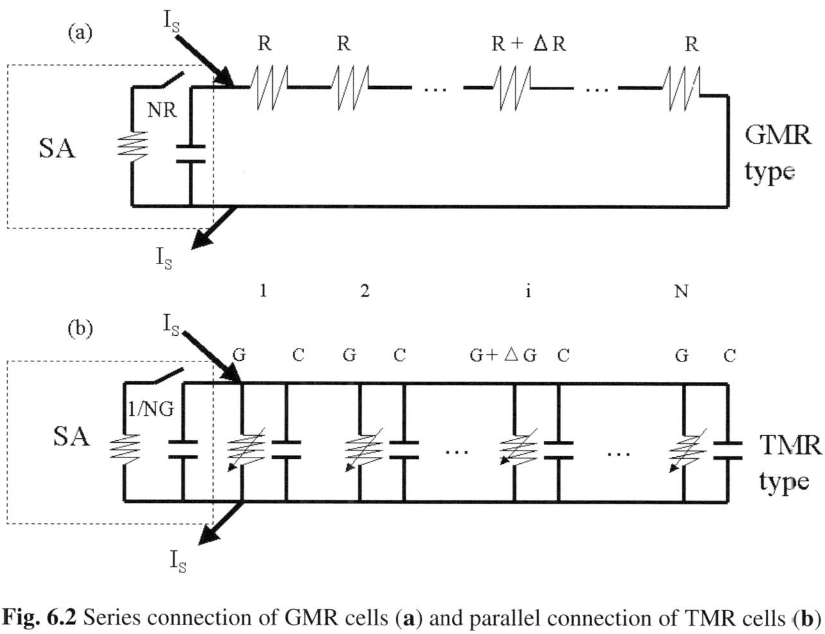

Fig. 6.2 Series connection of GMR cells (**a**) and parallel connection of TMR cells (**b**)

where F is a noise factor with theoretical value of unity [2]. This leads to the upper limit of N. MRAM using PSV has been developed up to a 1 Mb memory chip for satellite and military applications [3]. The chip employs dual redundancy, CMOS drive electronics with minimum gate lengths of 0.8 μm and a 5 V±10% power supply. Increased reliability was achieved by utilizing two memory cells per bit. The read cycle time was limited to 250 ns due to the small signal voltage of ±3 mV per cell during a read operation, which is attributed to the lower GMR of PSV. The small signal does not make a GMR type MRAM attractive for mainstream RAM applications.

6.1.3 MRAM Devices Using MTJ Cells

The MTJ cell functions very much like PSV devices using GMR sandwich materials, except the readout mechanism uses TMR rather than GMR. The MTJ cell is formed by two magnetic layers with different coercive forces, separated by a thin insulating layer as a tunnel barrier. Unlike PSV cells, which can use current through the cell to aid in switching, currents in the MTJ cells will be low due to their high impedance, and those low currents are not oriented properly to aid in switching. Therefore, MTJ cells use one strip line insulated from the cell as the drive line for switching. The MTJ cell has high impedance due to electron tunneling. These characteristics, in addition to higher MR than the GMR of SPV, can provide higher readout voltage with low sense current for MTJ cells. Recent advances in tunnel junction fabrication led to TMR changes of about 50% in 20–30 Oe field ranges and resistance-area products ranging from 10 $\Omega\mu m^2$ to 10^9 $\Omega\mu m^2$. Thus, TMR devices have an advantage over GMR devices for high–speed nonvolatile MRAM memory. In the following, MRAM using MTJ cells is described.

6.2 MRAM Architectures Using MTJ

A typical MRAM architecture using MTJ is based on MTJ memory elements integrated with MOS transistors [4,5] as shown in Fig. 6.3(a), A memory cell consists of a minimum sized active MOS transistor as the isolation device in conjunction with a MTJ to define the MRAM as shown in Fig. 6.3(b). The storing for an MTJ element can be achieved by using a bit line and a word line, where the sense ele c-tronics is common for an array of cells, and is gated to sense electronics through MOS transistors. The sense circuits for the MTJ memory do not require high current levels, thus greatly shortening the setting time required before a signal may be sensed. This would eliminate a substantial part of the read access time. For the design concepts in Fig. 6.3 the size of the MTJ devices is not critical because the transistor area is a much more significant part of the cell area.

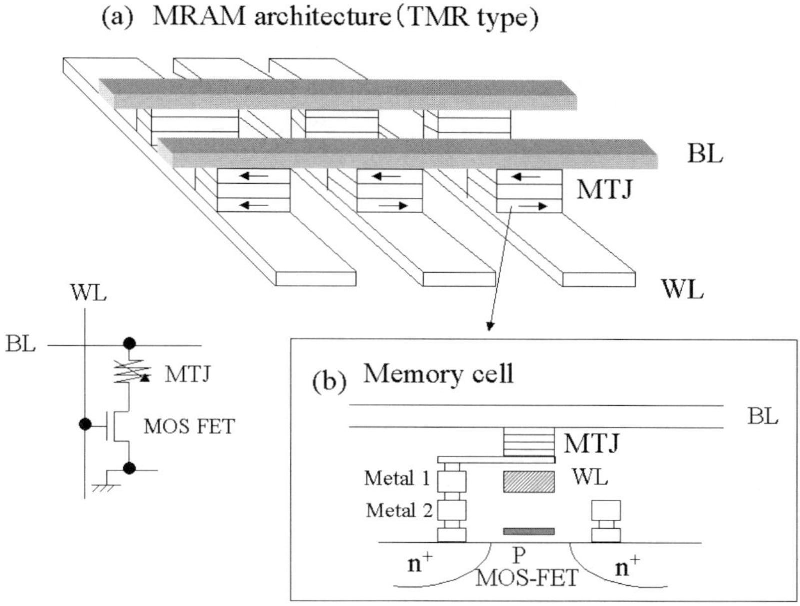

Fig. 6.3 MRAM architecture (**a**) and memory cell design (**b**) using a MOS transistor per MTJ cell

The write line is separate in a two–dimensional array, overlaying the array as shown in Fig. 6.3(b). These metal lines are conductors through which electrical current can pass. Typically magnetic memories use a combination of orthogonal bit and word lines in order to be able to individually address each magnetic memory cell to set or write its magnetic state. The vector combination of the orthogonal magnetic self-fields of the currents or current pulses passing through these lines is arranged such that the magnetic state of the selected memory element at the intersection of the chosen bit and word lines can be appropriately set. On the other hand the self-fields of these same currents must be such that the magnetic state of the half-selected devices along the same bit and word lines is not altered. Nevertheless, these latter cells will be magnetically distributed and it is very important that after many such disturbances the magnetic state of these cells does not creep either to some intermediate state or completely reverse. Unpatterned MTJs without exchange-bias layers appear to be susceptible to creep.

Another possible design of an MRAM device is shown in Fig. 6.4 [5,6]. Two perpendicular grids of lines make it possible to select an individual storage bit, located at each crossing. Each bit consists of a single MTJ, and a diode is connected in series with each junction for the purpose of being able to uniquely sense the resistance of a single element. This is a cross-point architecture which is the most desirable for a nanomemory device. The bit and word lines can be electrically biased so that only one diode is forward biased, thus the sense current flows only through a single chosen MTJ device in the memory array. The close proximity of the conductors to the MTJ reduces the write currents required. In addition the RC time constants on the array lines are determined by the capacitance of the diode rather than the high capacitance of the tunnel junction reducing sensing time compared to the parallel or matrix TMR arrangement. The cell requires a thin film diode integrated with VLSI metalization and MTJ processes which is a significant challenge [7].

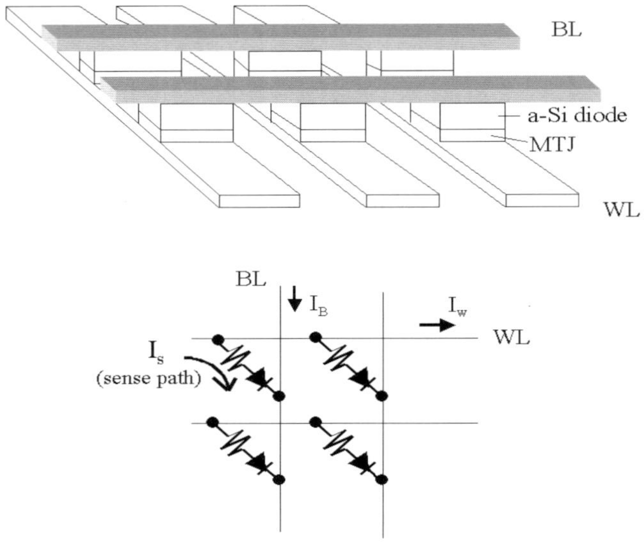

Fig. 6.4 MRAM architecture using a diode per MTJ cell

The ideality factor of the diode effects the differential resistance of the diode which must be smaller than the resistance of the MTJ element. This requirement implies that the IR drop across the TMR device must be greater than the ideality factor times the diode thermal voltage. The falloff of TMR to half its maximum occurs at a bias voltage in the range of hundreds of millivolts, therefore many times the thermal voltage of the diode. The requirement regarding the diode differential resistance therefore can be met without significant MR rolloff.

Since it provides isolation of the sense path from sneak paths through the unselected cells, the diode for the cross-point MTJ array must have an on to off conductance ratio in the range of four orders of magnitude for an array of 128 by 128 bits. The reverse bias on the unselected cells can be less than one volt. Conductance ratios of this range are achievable with thin film diodes. However, the on conductance of typical thin film materials is too low for cells with feature sizes less than a micron. Improved mobility thin film material would be required to implement cross-point MTJ cells.

6.3 Features of MRAM

The features of MRAM using MTJ are shown in Table 6.1, compared with those of other competitive memories. MRAM has the potential to be as fast as SRAM and as dense as DRAM with the additional advantage of nonvolatality. In principle, MRAM cells operate at low voltage, and the written two-spin states in MRAM cells are stable for small cell size, down to several tens of nanometers in length, below which the cells are unstable due to superparamagnetic properties of magnets. Thus, MRAM has potential as high density nonvolatile memory.

Table 6.1 Comparison of MRAM with competitive memories

	MRAM	FeRAM	FLASH	DRAM	SRAM
Nonvolatility	○	○	○	X	X
Read time	◎	○	○	○	◎
Write time	◎	○	△	○	◎
Endurance	◎	○	△	◎	◎
Cell size	◎	△	◎	○	△
Power consumption	△	◎	△	△	○
Soft error hardness	○	○	○	X	○

The writing power for MRAM, however, may be higher than that of FeRAM due to spin switching by the currents needed. FeRAM has a similar performance to MRAM: it has the properties of nonvolatilities and fast writing and reading. Thus, FeRAM is a powerful competitor to MRAM. It seems, however, that MRAM devices may be made denser, because the submicron devices using FeRAM may be difficult to be made from the point of endurance, which is limited up to 10^{12} repetitions. Flash memory requires longer time for writing, thus is essentially read–only memory (ROM). DRAM is volatile and requires large power consumption due to the refreshing operation needed. SRAM has a large cell size, since multiple MOS transistors are used.

6.4 Design of an MTJ Element

For the optimum device performance of MRAM using the MTJ the junctions are required to have optimum resistance as well as large TMR. Advances in tunneling junction fabrication led to TMR changes at RT up to 50% in 20–30 Oe field ranges and resistance-area products below $k\Omega\bullet\mu m^2$, promising MRAM. TMR in single tunneling junctions, however, typically decreases with increasing bias voltage across the junction and loses roughly 1/2 of the low voltage values at 300 – 400 mV, as shown in the previous section. A weak bias voltage dependence of the TMR is required for high-performance MRAM devices, because the voltage drop of several hundreds of mV across the junction is estimated for a way to sense an MTJ element by comparing voltage for the MRAM devices with a MTJ plus MOS transistor or diode switch architecture, which can provide a large signal-to-noise ratio.

The optimum junction resistance and minimum TMR ratio required for high-density and high-performance MRAM devices with an MTJ element plus MOS transistor switch architecture (1T1J) are discussed by taking account of the variations of the junction and MOS transistor resistances [8, 9]. This demands a large TMR at an elevated bias voltage and junction resistance of several tens of kΩ for MTJ elements. An important parameter for MTJ memory cell design is the *RC* time constant of the device as well as the signal across the cell. Consider a simple memory design using a MOS transistor per MTJ element as shown in Fig. 6.5. A constant current from the power source is supplied to the MTJ and MOS transistors through bit lines (BLs). When an MTJ element is selected by a MOS transistor switched on by a word line (WL) voltage, the voltage drop across the cell is detected by the sense amplifier connected to the BL, and the magnetic bit of "0" or "1" can be read by comparing the voltage with the reference voltage. The read time t_r of the memory device can be written as

$$(R + r_{BS})\, C_B = t_r, \tag{6.1}$$

where R, r_{BS}, and C_B are the resistance of the MTJ, the resistance of the MOS transistor chosen, and the capacitance of the BL, respectively. MOS transistors usually have a variation of the resistance $\beta = \Delta r_{BS}/r_{BS}$, which causes noise. Noise also arises from the variations of the junction resistance R and the resistance changes ΔR, which are denoted by γ and δ, respectively. Thus, the signal voltage V_s is given by

$$V_s \leqq V_{original} - V_{noise} = (\Delta R/2-(\beta r_{BS} + \gamma R + \delta\Delta R))\, I_s \tag{6.2}$$

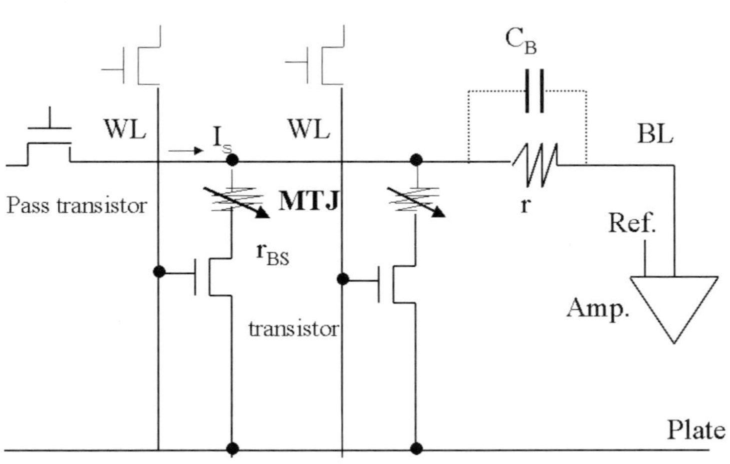

Fig. 6.5 An electrical circuit for the MRAM device with MOS transistor per MTJ cell

where I_s and ΔR are sense current and resistance change of the MTJ between two magnetic bits, respectively. Then, (6.1) and (6.2) lead to

$$\Delta R/R \geqq [2V_s + 2\,(\beta r_{BS} + \gamma R + \delta\Delta R\,)\,I_s] \,/\, (R\,I_s\,). \tag{6.3}$$

The bias voltage V_b is given by

$$V_b = R\,I_s \tag{6.4}$$

Equation (6.3) gives $\Delta R/R$ as a function of the MOS transistor parameters (r_{BS}, β), I_s, the junction resistance R and its variation γ and the signal voltage V_s. If we assume a signal voltage $V_s = 50$ mV, $r_{BS} = 1$ kΩ, $I_s = 10$ μA and $\beta = 0.2$, and neglect the variation of the TMR assuming $\gamma > \delta$, the TMR and $\Delta R/$ demanded can be obtained as a function of R as shown in Fig. 6.6, which exhibits for $\gamma = 0$ and ± 0.02. Considering the current status of TMR value the optimum junction resistance may be about 40 kΩ, which gives TMR = 25% and 29% at $V_s = 400$ mV for $\gamma = 0$ and ± 0.02, respectively. In this case the influence of the junction resistance variation is more serious than that of the variation of the MOS transistor, since the resistance is higher for the junction than for the transistor. The junction resistance of 40 kΩ requires junction • area products of 4 k$\Omega\mu$m^2 for submicron MTJ elements. The read time t_r is calculated to be 10 ns from (6.1) by assuming $C_B = 200$ fF for the circuit in Fig. 6.5.

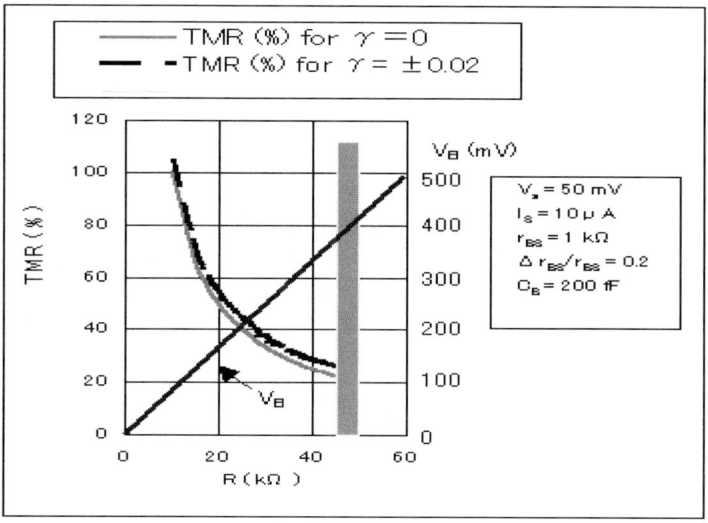

Fig. 6.6 TMR and bias voltage as a function of junction resistance estimated for the MRAM

The signal voltage V_s is proportional to TMR as shown in Fig. 6.7 exhibiting V_s as a function of TMR at $V_s = 400$ mV for the same condition as Fig.6.6 with $\gamma = 0$. For high-density and high-performance MRAM a higher signal voltage is desirable, requiring larger TMR at an elevated bias voltage, because there are other noise factors besides the variations of transistors and junctions considered here in the actual MRAM devices. Thus, double tunneling junctions described in the previous section may be desirable for high-density and high-performance MRAM, because TMR reduction due to bias voltage is significantly smaller than that of single junctions. In fact, the voltage change across the junction divided by 2, $\Delta V/2$ between the two magnetic bits, which corresponds to signal voltage for the 1T1J architecture, is significantly larger for double junctions than single junctions, of which an example is demonstrated in Fig. 6.8 for the junctions shown in Fig. 5.22. The maximum value is approximately 100 mV for the double junction, while it is 50 mV for the single junction. This result is mainly attributed to the difference of the bias voltage dependence of TMR.

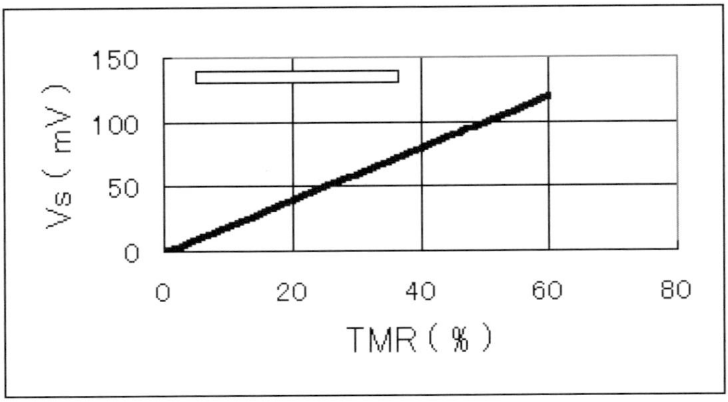

Fig. 6.7 Signal voltage V_s vs TMR for the MRAM circuit in Fig. 6.5

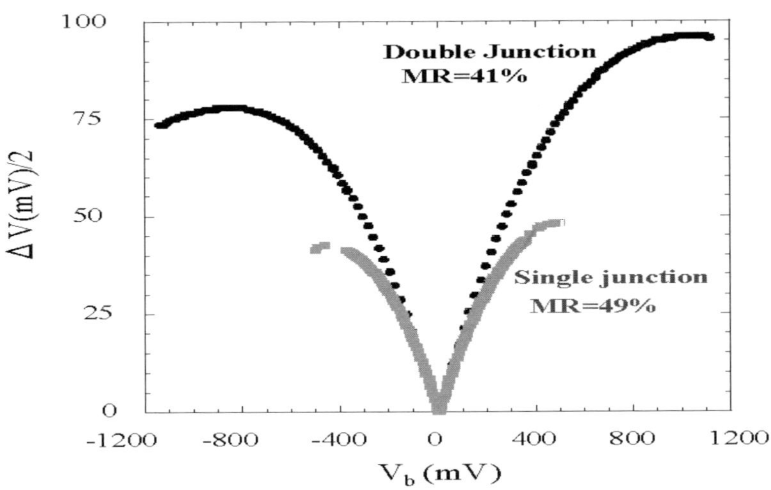

Fig. 6.8 Signal voltage vs bias voltage for single and double junctions

6.5 Spin Switching Field for MTJ Elements

6.5.1 Single–Domain Calculation

Spin switching of a spin-valve type MTJ element is discussed in the single domain coherent rotation model [10]. Assuming the magnetic moment of the free magnetic layer makes an angle of θ with the length of the cell as illustrated in Fig. 6.9, the demagnetization energy density is

$$2\pi M^2\,(T/W)\,\sin 2\theta\,, \tag{6.5}$$

where M is the magnetization of the free magnetic layer, T is its thickness, and W is its width. The energy density from the applied magnetic field H_a for switching the magnetization to the opposite direction is $H_a\,M\cos\theta$. The restoring torque density of the demagnetization energy is obtained by differentiating the demagnetization energy density with θ. It is

$$4\pi M^2\,(T/W)\,\cos\theta\,\sin\theta. \tag{6.6}$$

The maximum absolute value of the restoring torque density occurs at a value of θ equal to 45°. Thus, the maximum demagnetization torque density which is retarding the rotation of the magnetization in the opposite direction is

$$2\pi M^2\,(T/W). \tag{6.7}$$

The torque density from the applied field is

$$-H_a\,M\sin\theta. \tag{6.8}$$

Fig. 6.9 Sketch of the momory cell along with magnetization, demagnetization field and applied magnetic field

This torque density will increase in absolute magnitude as the angle θ increases from 0° to 90°, while the demagnetization retarding torque density has a maximum at 45°, then decreases. If we consider submicron devices, the magnetic anisotropy energy is much smaller than the demagnetization energy, thus it can be neglected. The magnetization should switch when θ is 45°. After switching, the magnetization aligns with the cell length in opposite direction because both the applied field and the demagnetization field favor this direction. The torque density of the applied magnetic field at $\theta = 45°$ is

$$-H_a M / \sqrt{2}. \tag{6.9}$$

The sum of (6.7) and (6.9) is equal to zero at the switching transition angle of 45°. This yields the switching field for the free magnetic layer

$$H_a = 2\sqrt{2}\pi M T / W. \tag{6.10}$$

This suggests a $1/W$ dependence of the switching field, explaining experiment [10]. Thus, in order to obtain a low switching field for writing, which is very important for actualizing MRAM with low power consumption, the memory layer thickness should be as thin as possible. Furthermore, (6.10) states that the MRAM consumes a larger writing power for higher bit density due to the smaller width.

6.5.2 Micromagnetic Simulation

A micromagnetic model using the Landau-Lifshiz-Gilbert (LLG) equation has been used to investigate the switching field variation in patterned film elements as well as dynamical switching behavior. A patterned film element is modeled by discretizing it into a two-dimensional array of tetragonal cells. To study the magnetic switching properties, an external field is always applied along the length direction of the film element in the film plane. The amplitude of the field is varied stepwise with a 2 Oe field step. Integration of the LLG equation is performed after each field change until equilibrium is reached. We now describe the correlation of the magnetic switching characteristic with the element dimension and the intrinsic magnetic anisotropy properties.

A patterned pseudo-spin-valve NiFeCo/Cu/NiFeCo trilayer element was studied [11]. Each NiFeCo layer is discretized into an array of tetragonal cells with a dimension of 10 nmx10 nm x film thickness for each cell. The saturation magnetization of NiFeCo is taken to be 860 emu/cm^3. A 2.5 Oe uniaxial anisotropy field is assumed to be along the element length direction for both NiFeCo layers. The thicknesses of the two NiFeCo layers are different so that the magnetization of the two layers switch at two distinctly different field magnitudes. The thinner layer, referred to as the soft layer, switches at a smaller field magnitude while the thicker layer, referred to as the hard layer, switches at a significantly higher field value. Here, the two NiFeCo layer thicknesses are assumed to be 1 nm and 2 nm, respectively. The Cu interlayer is 3.0 nm thick, which leads to negligible interlayer exchange coupling. So, the only interlayer coupling considered is magnetostatic interaction in the calculation. This assumption can also be applied for MTJ. An external field was applied along the magnetic easy axis.

Figure 6.10(a) shows a calculated GMR transfer curve for an element with a length of 1 μm and a width of 0.2 μm [11]. The field magnitude for soft layer switching back is 50 Oe while the switching field of the hard layer is 80 Oe. The difference between these switching fields is not sufficient in practice. This phenomenon is more pronounced as the width of the memory element is reduced.

Figure 6.10(b) shows the calculated results for 0.1 μm element width [12]. As the width of the element is reduced, the soft layer switching back field significantly increases, thereby requiring a larger read field. The large soft layer switching back field is due to the interlayer magnetostatic interaction, which induces an unstable parallel state and asymmetric switching fields.

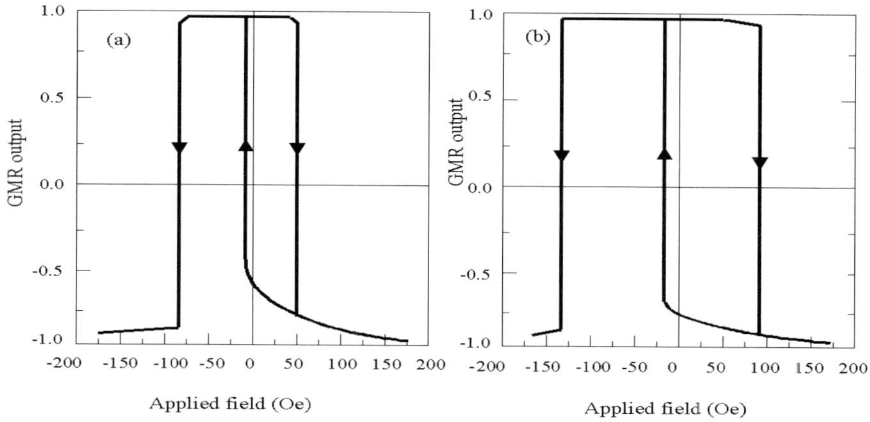

Fig. 6.10 Simulated GMR transfer curves of (**a**) 1 μm x 0.2 μm and (**b**) 1 μm x 0.1 μm tri-layer structure memory cells

In order to eliminate the interlayer magnetostatic fields, a synthetic antiferromagnet (SyAF) can be used to replace the pinned layer. A micromagnetic model using LLG equations for multilayer structured GMR film has been developed. For the results presented here, the modeled SyAF biased spin valve is Co(1nm)/Ru(0.7nm)/Co(1nm)/Cu(3nm)/NiFeCo(3nm)[12]. The interlayer exchange coupling between the two Co layers is 1,000 Oe. The free layer of the spin valve is a NiFeCo alloy film with a saturation magnetization of 1,000 emu/cm^3 and induced uniaxial anisotropy field of 5 Oe in the direction of the spatially uniform applied field. For the SyAF biased spin valve of a square element with 0.5 x 0.5 μm^2, when the external field was applied in the same direction of the free layer, the magnetization of the top AL layer was initiated to be parallel to that of the free layer, forming the low resistance state of the spin valve. Should the SyAF magnetization configuration be rigid, the parallel state would have been unchanged under the applied field. However, when the field magnitude reached 40 Oe, the SyAF axis rotated to the direction orthogonal to the applied field, then immediately flipped over.

For the case of elongated elements the effect of element shape anisotropy for constraining the magnetization of the SyAF structure was investigated [12]. Figure 6.11(a) shows a calculated GMR hysteresis curve for a 0.5 x 0.25 μm^2 element with the external field applied in the length direction [12]. The irreversible switching field of the SyAF axis (Point A in Fig. 6.11(a)) became almost twice that of the free layer. The risk of relying only on the shape anisotropy is the possibility of multidomain formation in the SyAF layer. Since the irreversible magnetization flipping of the SyAF layer is driven by the differential field, applying an exchange pinning field to the bottom SyAF layer can be effective for preventing the flipping. Figure 6.11(b) shows the calculated GMR output for an SyAF biased spin valve with an additional antiferromagnetic layer for providing the exchange pinned field to the bottom SyAF layer, schematically shown as the inset [12]. The memory element has a dimension of 0.5 x 0.25 μm^2 The GMR hysteresis, calculated with 50 Oe for the AF/F exchange pinned field, shows nearly ideal performance. The AF/F exchange pinning can prevent multidomain formation in the SyAF structure, which can easily occur in either elongated or square elements.

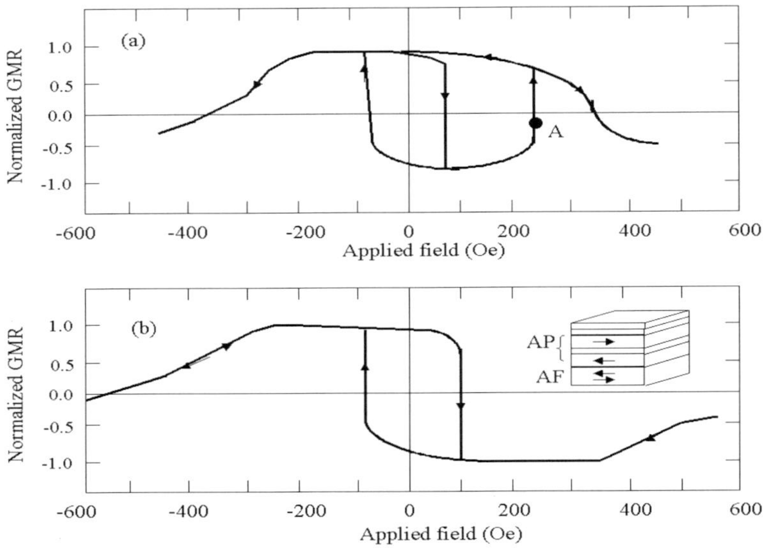

Fig. 6.11 Calculated GMR hysteresis curves of (**a**) an elongated element and (**b**) an elongated element with 50 Oe AF/F pinned field on the bottom AP layer

6.5.3 Two–Dimensional Magnetic Switching

Memory cells in MRAM devices require magnetic fields to be applied in two dimensions, thus it is important to study the response of tunneling junctions to fields applied along both the easy- and hard-axis directions. If the Stoner-Wohlfarth (SW) model, which is coherent rotation of the magnetization, is assumed, the switching field for the magnetic fields H_x and H_y applied in two dimensions, is given by the points on the curve given by the relation $H_x^{2/3} + H_y^{2/3} = H_k^{2/3}$, which is known as the asteroid curve and is shown in Fig. 6.12. H_k is the anisotropy field of the memory layer of the MTJ and is a function of the memory element dimension. When the two fields are such as to give points outside the asteroid, switching occurs by coherent rotation. In this section the properties of micron-size MTJs subject to two-dimensional magnetic fields in the plane of the device are measured and compared with the asteroid curve [12].

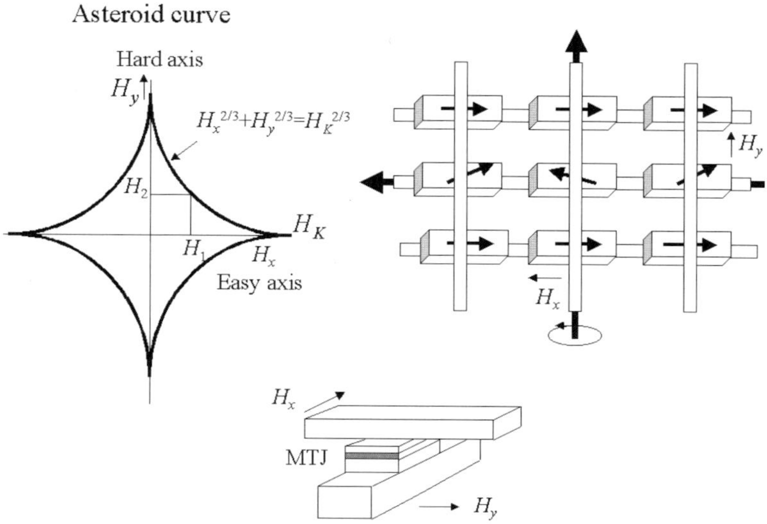

Fig. 6.12 Stoner–Wohlfarth critical curve (asteroid curve) for an ideal single-domain particle with uniaxial anisotropy

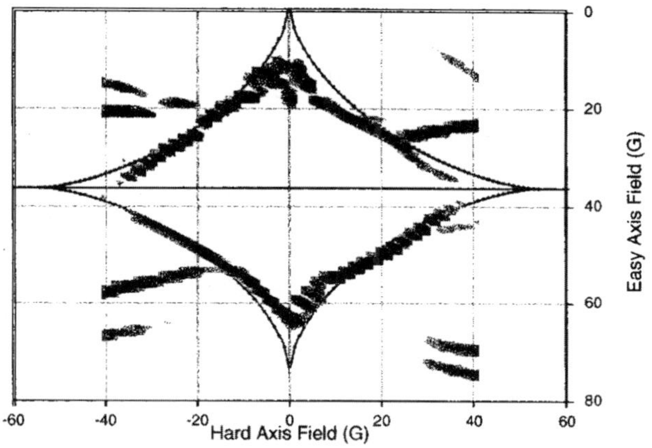

Fig. 6.13 Sample experimental critical asteroid curve. The darker region indicates a higher probability of switching. The solid black line is a theoretical SW fit to the data.

The sample layer sequence is Si(100) substrate/5 nm Ta/25nm Al/4nm $Ni_{60}Fe_{40}$/ 10 nm FeMn/6 nm Co/0.7 nm Ru/3 nm Co/1.1 nm Al_2O_3/7.5 nm $Ni_{60}Fe_{40}$/25 nm Al/7.5 nm Ta. The Ru and FeMn layers are inserted to control the magnetic properties of the pinned layers. The TMR was measured on a probe station equipped with magnets providing a 2D field. For each sample hundreds of MR hysteresis loops along the easy axis were recorded at a fixed hard axis. The hard axis field was then incremented and the process repeated. For each MR loop, two switching fields were recorded. These points were plotted in a 2D field space to give the asteroid curve. One such curve is shown in Fig. 6.13 [13]. While the SW model predicts a curve centered at the origin of field space, the actual asteroids are generally offset from the origin in both directions. In addition, experimental asteroids tend to be longer along the hard axis than the easy axis, different from the SW model prediction of equal critical fields along both directions. The offset of the origin of the critical curve is caused by magnetic interactions.

Two mechanisms contribute to this offset. The first is the uncompensated magnetic poles at both ends of the pinned layer creating a nonuniform field in the free layer. This magnetostatic coupling appears to scale roughly inversely with the length of the pinned layer. This coupling is expressed as

$$H_M = B / L, \tag{6.11}$$

with B an adjustable constant depending on the sample structure and L the length of the pinned layer. The second is the Neel "orange peel" coupling through the tunneling barrier and due to the roughness of the interfaces. Independent of sample dimensions, this affects several material parameters as

$$H_N = \pi^2/\sqrt{2}\,(h^2/\lambda t_F) Ms \exp(-2\,\pi\sqrt{2} t_s/\lambda), \tag{6.12}$$

where h and λ are the amplitude and wavelength of the interface roughness, t_s and t_F the thicknesses of the tunneling barrier and free layer, respectively, and H_s is the saturation magnetization of the pinned layer. Equation (6.12) assumes a sinusoidal interface roughness as an approximation. The net offset field is $H_{off} = H_M + H_N$.

The SW model predicts an asteroid which is symmetric along the easy- and hard-axis field directions and whose area scale as $(2H_k/M_s)^2$. For rectangular samples with a high aspect ratio, H_k due to shape anisotropy scales inversely with the sample width (w):

$$H_k = 4\pi M_s t_F / w \tag{6.13}$$

Therefore, the area of the critical curve scales as w^{-2}. The areas of critical curves measured for many samples with varying aspect ratios are plotted versus junction width with a log-log scale in Fig. 6.14 [13]. The area scales as a power law in the junction width, *ie.*, $S = C\,w^{-1.4}$, where C is a constant which depends only on the sample aspect ratio. Samples with differing aspect ratios are seen to lie on parallel lines. The discrepancy between the theoretical power law of w^{-2} and the experimental $w^{-1.4}$ is probably caused by the multidomain structure in the samples.

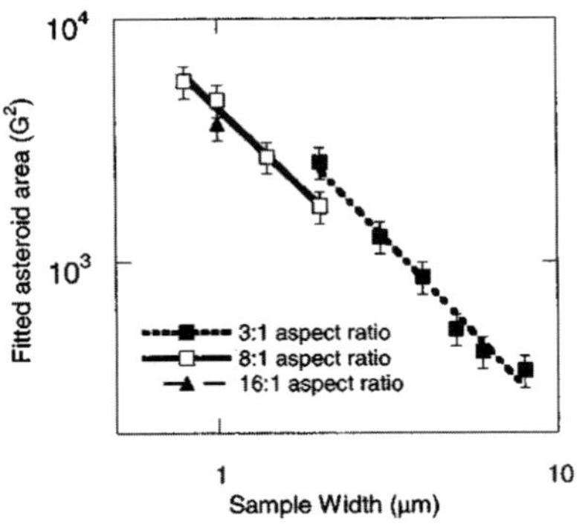

Fig. 6.14 Asteroid area (S) as a function of sample width. The two solid lines are power-law fits ($S = Cw^{-1.4}$) to the data for rectangular junctions with aspect ratios 3:1 and 8:1

6.6 Problems in Fabricating MRAM

6.6.1 Integrated Circuit Compatibility [2]

There are two aspects to IC compatibility: circuit compatibility and process compatibility. Integrated circuit impedance levels are intrinsically much closer to those of TMR devices than GMR devices, especially in the case of MOS- and diode-based circuits. It is anticipated that integration of TMR devices with silicon circuits will introduce some challenges, including:

1. Making reliable contacts to the electrodes without damaging the barrier during completion of integrated circuit processing.
2. Developing compatible metallurgies between the MTJ and the circuit.
3. Preparing the surface quality of the IC substrates suitable for MTJ deposition. The smoothness of dielectrics used for IC is not an issue for the circuit, but could be a very large one for TMR device operation.
4. The temperature durability of the TMR devices must allow for an annealing temperature of about 673 K to anneal out radiation damage from transistors caused by plasma processes.

6.6.2 Uniformity of MTJ Elements

Uniformity of the magnetoresistance and absolute junction resistance of the MTJ cells are critical in the 1T1J architecture, since the absolute value of the MTJ resistance is compared with a reference cell during read mode. If the active device resistances in a block of memory show a large variation, a signal error can occur when they are compared with the reference cell. The resistance of the MTJ cell is exponentially dependent on the thickness of the Al_2O_3 barrier. Therefore, it is anticipated that small variations in the Al_2O_3 thickness would result in large variations in the resistance. Current technology in junction fabrication makes feasible to prepare very uniform resistance and signal distributions across a six inch wafer and wafer-to-wafer [5]. Uniformity of the switching field for MTJ cells is also critical, since the writing power is dependent on the switching field. Parkin showed resistance versus field curves for two sets of 14 nominally identical junctions [4] with nominal dimensions of 2 x 4 and 0.6 x 1.2 $(\mu m)^2$. The easy axes of the junctions are along their longer axis in both cases. The variation in magnetic switching field from junction to junction is larger, in particular, in the smaller junctions, because of the larger demagnetization fields. The larger variation in switching field is most likely attributable to small defects in the dimension of the shape of the MTJ elements.

6.7 Current Trial of MRAM Fabrication

Recently, MRAM was fabricated using MTJ integrated with CMOS technology by IBM and Motorola [14, 15]. The latter fabricated a 512 bit MRAM device with the 1T1J architecture using reference cells. The former, on the other hand, succeeded in making 1 kbit MRAM using 0.25 μm CMOS technology and the twin cell arrangement where two adjacent cells are always written to opposite states as shown in Fig. 6.15. The twin cell arrangement doubles the signal available for sensing, compares resistances of adjacent identical MTJs and provides protection from common-mode noise as compared to sensing the cell resistance relative to the reference resistance (1T1J). However, the twin cell arrangement leads to a larger cell size, thus is not suited for high-density MRAM. The sense line swing was limited to 300 mV in the IBM device to achieve high performance and to avoid the decrease of TMR that occurs at higher voltage.

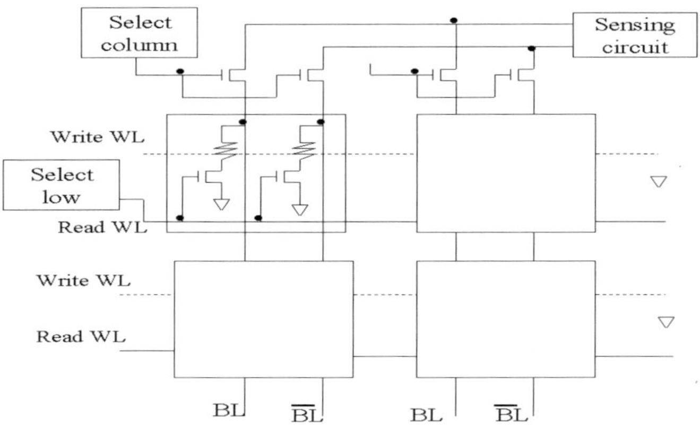

Fig. 6.15 Memory array circuit with twin cell arrangement

Table 6.2 Evaluated results for memory array circuit with twin cell arrangement.

	IBM	Motorola
TMR(%) / Resistance (kΩ)	18 / 2.5	26 / 23
TMR element size (μm^2)	0.4×1.2	0.9×1.5
Memory capacity (bits)	1K (twin cell)	512 (256x2) (single cell)
Design rule F (μm)	0.25	0.6
Writing time (ns)	10	24
Reading time (ns)	10*	34
Cell area (μm^2)	3.07 (49 F^2)	7.1(20 F^2)
Tip area (mm^2)	1.5	—
Voltage(V)	2.5	3
Consumption power (mW)		
Read	5	2.4
Write (pulse width)	40 (2.5 ns)	48 (5 ns)

* 3 ns for TMR = 25%

The evaluated results are shown in Table 6.2, in addition to the properties of the MTJ materials and the cell size used. Nonvolatility and the high access time of below 10 ns were demonstrated in the 1 Kb memory array by IBM. The read and write on-chip power at 2.5 V and 100 MHz were 5 mW and 40 mW, respectively. The power of 40 mW corresponds to 16 mA for the writing current, and should be reduced for actualizing the MRAM. The cell size is 49F ^{2}, where F is the design rule, and is significantly large due to the twin cell structure.

Motorola further reported a 256Kb MRAM using 1T1J in the ISSCC conference, February, 2001. The access time for both writing and reading is 35 ns and the power consumption for reading and writing is 24 mW and 528 mW, respectively.

6.8 Scalability of MRAM: Challenges for High Bit Density

Magnetic tunneling junctions can possess stable magnetic bits down to submicrons keeping a large TMR without degrading, by microfabrication. Thus, MRAM has the potential for creating very large memory capacity over gigabits. In order to advance with this goal, however, there are some challenges for scalability to be overcome.

6.8.1 Resistance

For obtaining high-density MRAM with small cell sizes the junction resistance *R* must be adjusted to the CMOS circuit. The product of junction resistance (*R*) x junction area (*A*) should be kept constant, while *R* increases with decreasing *A*. Thus, the resistivity of the junction must be reduced by decreasing the cell size. This can be accomplished by reducing the barrier thickness as shown in Fig. 5.11. For the demonstrated MRAM the product *RA* is several tens of $k\Omega\mu m^2$. If we consider the cell size down to $0.01\mu m^2$, *RA* should be reduced down to 0.1 $k\Omega\mu m^2$. Figure 6.16 shows TMR as a function of *RA* for $IrMn/CoFe/AlO_x/NiFe$ junctions [16], exhibiting large TMR down to 0.1 $k\Omega\mu m^2$, although the TMR decreases with decreasing *RA*. This means the junction resistance is not an obstacle for MRAM scaling, although further study is desirable to obtain junction materials with larger TMR and smaller resistance.

6.8.2 Signal Voltage

A large signal voltage such as 100 mV is desirable for high-density MRAM. Quite a large TMR above 100% at a low bias voltage may be needed to surmount the requirement using single tunneling junctions. A solution for that is to prepare the junctions using half metallic ferromagnets possessing a spin polarization of 1. However, it has not yet been possible to achieve a large TMR at RT for the junctions, although a large TMR of 450% has been obtained at low temperatures [17].

Some of the half metals have a low Curie temperature, which is one of the causes of the low TMR at RT. Another cause may be that the band structure is very sensitive to the surface states and the composition of half metals. Since tunneling is extremely surface sensitive, *in situ* study, such as Auger or XPS, is necessary to examine the surface quality and optimize the deposition of half metals.

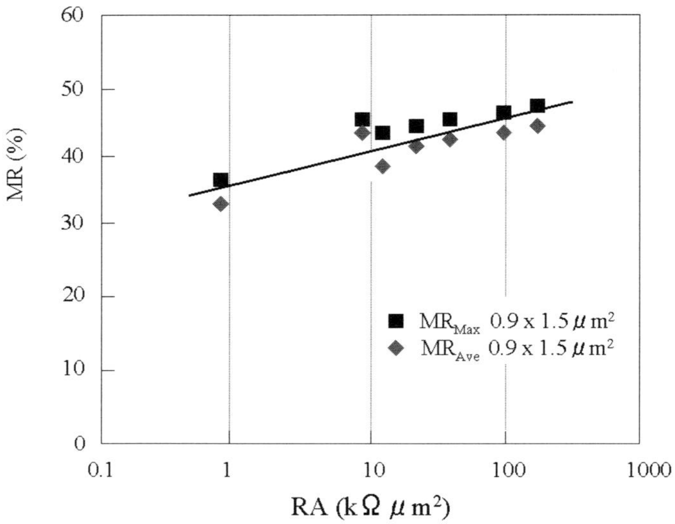

Fig. 6.16 MR vs RA for a IrMn/CoFe/AlO_x/NiFe tunnel junction patterned to 0.9 x 1.5 μm^2

Another solution for obtaining the large signal voltage is double tunneling junctions, since they have a smaller bias voltage dependence of the TMR, as demonstrated in Sect. 6.4. Additional tunneling such as spin-dependent resonant tunneling in double tunneling junctions can enhance the TMR [18, 19]. It will be significant that the central ferromagnet is made with a very thin layered structure and small roughness to observe such an enhancement. Thirdly, new architectures which can enhance the signal voltage (such as the twin cell structure) and admit the small cell size need to be explored.

6.8.3 Writing Current

The switching field needed for writing becomes large with decreasing MRAM element size as described in Sect. 6.5. This requires a large field from the writing current lines, which depends on the conversion efficiency of the current to the field due to the line arrangement. In order to obtain the large field from the straight line current MRAM elements should be positioned along the lines as nearly as possible. Figure 6.17 shows the magnetic field obtained from the current as a function of the line width for various separations δ between the bit and the line [16]. This suggests that the field steeply increases for small δ. The important thing is that the responsibility of the bit for the hard axis field for small bit sizes.

Figure 6.18 exhibits the normalized asteroid curves for three bit widths with an aspect ratio of 2 [16]. It shows that the bit with the smallest width responds well to the hard axis field. Further study is required for investigating magnetic coupling between the current line and the bit.

Recently, an effective method of current-induced switching of magnetic orientation was theoretically predicted [20] and observed experimentally in nano-sized junctions [21]. The switching can be understood in terms of an exchange of angular momentum between the spin-polarized current and a magnetic cluster situated between the electrodes. About 100 μA of the switching current has been estimated for an NiFe particle with 30 nm in diameter. This technique may be useful for double tunneling junctions with small size, thus high bit density MRAM.

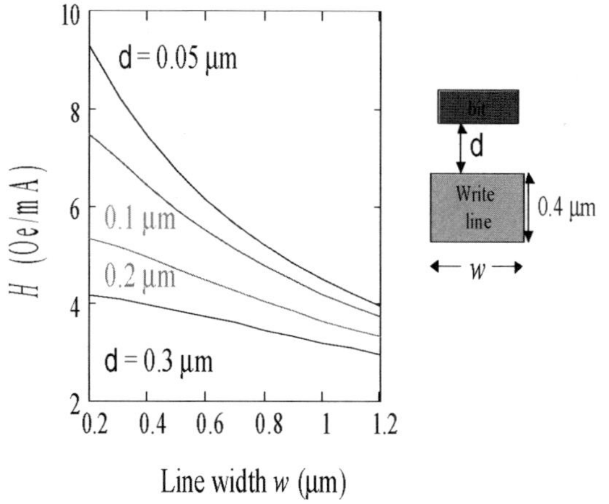

Fig.6.17 Magnetic field obtained from the current as a function of the line width for various separations δ between the bit and the line

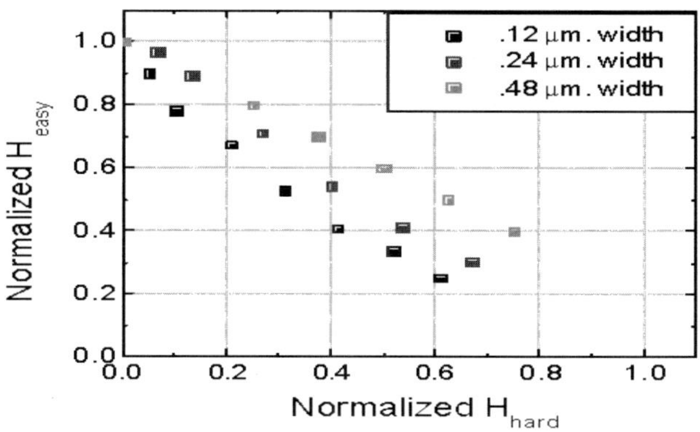

Fig.6.18 Normalized asteroid curves for three bit widths with an aspect ratio of 2

6.8.4 Superparamagnetism

The effective energy barrier for a switching field becomes lower with decreasing MRAM bit size. If a uniform switching mode is assumed, the switching field is given by $E = K_u V$, where K_u and V are the uniaxial magnetic anisotropy energy and bit volume, respectively. Thermal fluctuation becomes effective for small bit size, and the switching depends on time such as $\tau = \tau_0 e^{\alpha}$, where $\alpha = K_u V/k_b T$ and $\tau_0 \sim 10^{-9}$ s. Thus, τ exponentially depends on α and the magnetization of the bit is stable for α larger than a critical value. This means that the bit volume and switching field of the MRAM must be optimized for reduced bit size. It is known that MRAM does not exhibit superparamagnetism down to 0.1 μm. Research is required for a smaller bit size than 0.1 μm .

Finally, the load map of MRAM, by taking account of the announcement by some companies is shown in Fig. 6.19, compared with that of FeRAM. The trend of the feature size F of a MOS transistor is also shown in Fig. 6.19. IBM and Infineon Technologies AG announced that they will collaborate on MRAM, and an MRAM product with 265 Mb could be available commercially as soon as 2004.

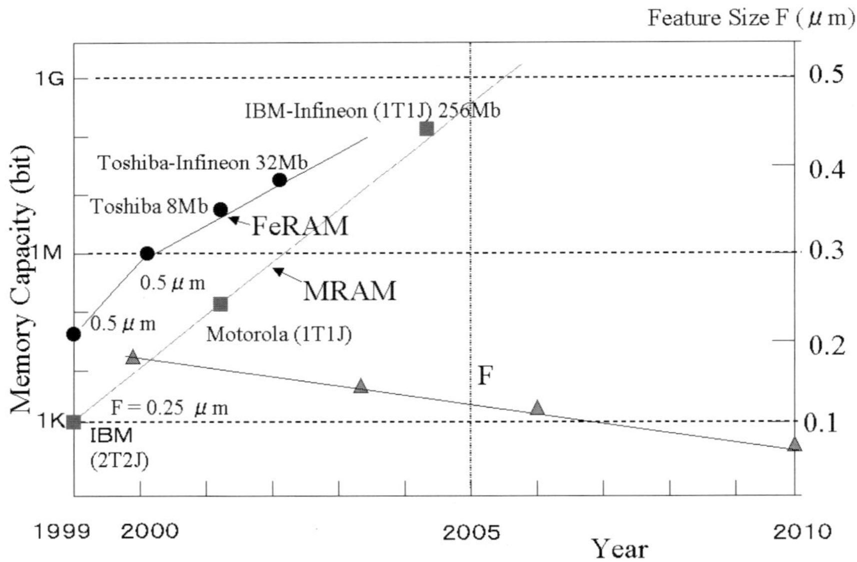

Fig. 6.19 Load map of MRAM compared with that of FeRAM. The trend of the feature size *F* of the MOS transistor is also shown

References

[1] J. M. Dauton, *Thin Solid films*, **216**, 162 (1992).
[2] J. M. Dauton, *J. Appl. Phys.* **81**, 3758 (1997).
[3] J. L. Brown and A. V. Pohm, *IEEE Trans. on Components, Packaging and Manufacturing Tech.* Part A, **17**, 373 (1994).
[4] S. S. P. Parkin, K. P. Roche, M. G. Samant, P. M. Rice, R. B. Beyers, R. E. Scheurlein, E. J. O'Sullivan, S. L. Brown, J. Bucchigano, D. W. Abraham, Yu Lu M. Rooks, P.L. Trouilloud, R. A. Wanner and W. J. Gallagher, J. Appl. Phys. **85**, 5828 (1999).
[5] S. Tehrani, E. Chen, M. Durlam, M. DeHerrera, J. M. Slaughter J. Shi and G. Kerszykowski: *J. Appl. Phys.* **85**, 5822 (1999).
[6] R.E. Scheuerlein, *Proc. of Int'l. Non Volatile Memory Technology Conf.*, P47 (1998).
[7] R. C. Sousa, P. P. Freitas, V. Chu and J. P. Conde, *Appl. Phys. Lett.* **74**, 3893 (1999).
[8] K. Inomata, *J. Mag Soc. of Jpn.* **23**, 1826 (1999).
[9] K. Inomata and Y. Saito, *J. Appl. Phys.* **87**, 6064 (2000).
[10] E. Y. Chen, S. Tehrani, T. Zu, M. Durlam and H. Goronkin, *J. Appl. Phys.* **81**, 3992 (1997).
[11] Y. Zheng and J. G. Zuh, *IEEE Trans. Magn.* **33**, 3286 (1997).
[12] J. Z. Zhu and Y. Zheng, *IEEE Trans. Magn.* **34**, 1063 (1998).
[13] A. Anguelouch, B. D. Schrag, G. Xiao, Yu Lu, P. L. Trouilloud, R. A. Wanner, W. J. Gallagher and S. S. P. Parkin: *Appl. Phys. Lett.* **76**, 622 (2000).
[14] R. Scheuerlein, W. Gallagher, S. S. P. Parkin, A. Lee, S. Ray, R. Robertazzi and W. Reohr: *Proc. of ISSCC* 2000 (2000)
[15] M. Durlam, P. Naji, M. DeHerrera, S. Tehrani, G. Kerszykowski and K. Kyler: *Proc. of ISSCC* 2000 (2000)
[16] S. Tehrani: Private communication.
[17] M. Viret, M. Drouet, J. Nassar, J. P. Contour, C. Fermon and A. Fert: *Europhys. Lett.* **39**, 545 (1997).
[18] X. Zhang and B. Z. Li: *Phys. Rev.* **B56**, 5484 (1997).
[19] T. Kishi and K. Inomata: *J. Magn. Soc. Jpn.* **21**, 501 (1997).
[20] J. C. Sloncwesky: *J. Magn. Magn. Mater.* **159**, L1 (1996).
[21] J. Z. Sun: *J. Magn. Magn. Mater.* **202**, 157 (1999).

7 Other GMR Devices

7.1 Bipolar Spin Transistor

7.1.1 Spin Injection

Consider steady state current flow in the circuit depicted in Fig. 7.1 [1], where two identical strong ferromagnetic layers (F_1 and F_2) are arranged with a distance on a thin nonmagnetic metal (N) layer. If F_i (i = 1, 2) is uniformly magnetized and the magnetization $\boldsymbol{M}_i$ is strongly oriented, it is a property of F that the current carrying electrons have spin down. When the battery drives a circulation, the current crossing the F_1-N interface has a net spin polarization and each electron remains polarized for a mean time T_2. In the diffusive conduction process, each electron is constantly scattered so that its mean free path $l = v_F\tau$ is quite short, where v_F and τ are the Fermi velocity and the mean scattering time, respectively. The macroscopic current results from a net drift of electrons, when each electron nudges the one next to it. But only a small fraction, about 1 in 1000 (*ie.*, τ/T_2 ~0.001), of these scattering events alters the spin state of the electron. The spin state of a down-spin electron eventually becomes randomized as an admixture of up-and down-spin

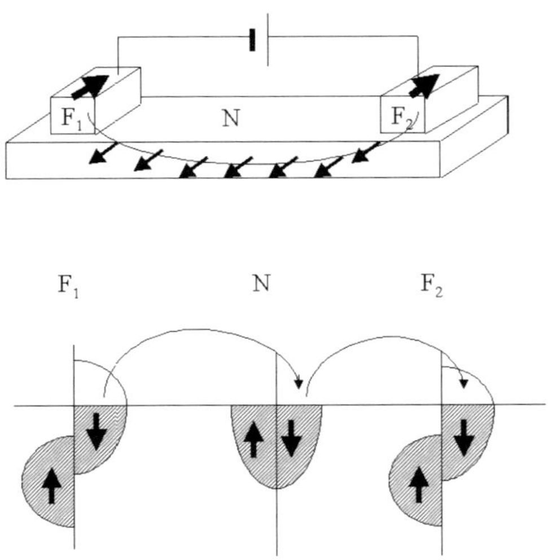

Fig. 7.1 Injection of spin polarized current into a paramagnetic bar from ferromagnetic pads

states, so that the spin polarization injected at the F_1-N interface decays away exponentially with a length scale called the spin diffusion length $\delta_s \sim (1000)^{1/2}\, l$.

The spin injection experiment measured $\leftarrow \delta_s$ directly by using a second F_2 film as shown in Fig. 7.1 as a detector of spin polarization electrons. By measuring the spin polarized current of several samples of varying length *L*, nonmagnetic metal thickness, the value $\delta_s = 0.5$ mm was determined for pure bulk Al at cryogenic temperatures [2]. The spin lifetime T_2 was independently measured to be $T_2 \sim 10$ ns by employing a transverse magnetic field to dephase the spins. The relationship [3] $\delta_s = v_F\, (\tau\, T_2\, /3)^{1/2}$ was thus verified, validating their model of spin polarized transport in ferromagnetic-nonmagnetic metal systems. In metal films spin diffusion lengths are of order 1 to 10 μm and spin relaxation times are 10 ps to 1 ns.

7.1.2 Spin Transistor Operation

7.1.2.1 Thermodynamic Description

The spin transistor is a trilayer, three terminal device as shown in Fig. 7.2 [4] which is similar to a semiconductor bipolar transistor. The emitter and collector layers of the spin transistor are ferromagnetic films characterized by a magnetization $\boldsymbol{M}_i$ (i = e or c) that lies in the plane of the films and has an externally manipulated direction. The base layer is a nonmagnetic metal such as gold, copper or silver.

First consider an F_1-N bilayer where the thickness of N is much smaller than δ_s. The electric current crossing the F_1-N interface, the emitter current I_e is spin polarized; F_1(the emitter) provides to N a current of oriented dipoles at the rate $I_M = \beta I_e/e$, where I_e/e is the number of current of carriers, *e* the electronic charge and β the Bohr magneton. Once in N, polarized dipoles lose their ordered orientation after $1/T_2$. The resulting nonequilibrium distribution of spin polarized electrons, equivalently the nonequilibrium magnetization *M*, is determined by a balance between the rates of injection and relaxation, $M = I_M T_2/Ad$, where *Ad* is the volume that is occupied by the nonequilibrium spins with *A* the area of interfacial contact and *d* the nonmagnetic layer thickness. The key to spin transistor action is the non-

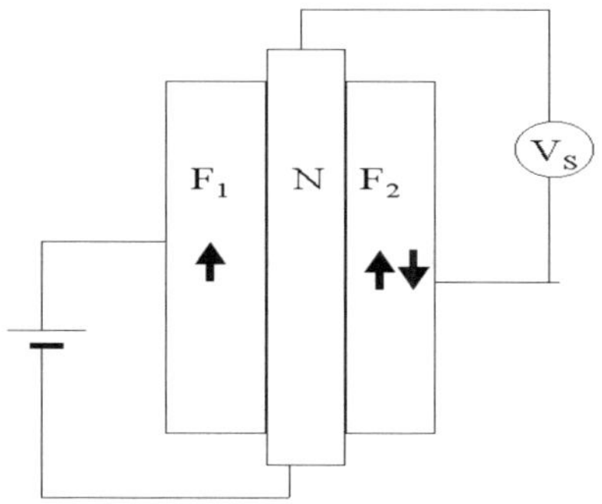

Fig. 7.2 Schematic drawing of spin transistor device.

equilibrium distribution M. It is a highly ordered state, therefore low in entropy, and the low entropy is provided at a cost of energy, $U_M = \beta\, M/\chi$ (with χ the Pauli susceptibility), called an effective Zeeman energy. It represents the magnetic energy of each spin polarized electron resulting from its interaction with the effective magnetic field (M/χ) associated with all of the other spin polarized electrons. This energy is capable of generating current flow.

A three-terminal device can be made connecting the back of F_2 (the collector) to a detector (either a current meter, a voltmeter, or an arbitrary load) and then to the ground of the battery. When the orientation M_c is parallel to M_e, F_2 acts as a reservoir of up-spin electrons. Each electron in the nonequilibrium spin distribution in N has excess energy U_M, where the spin population in F_2 is an equilibrium distribution, held in equilibrium by the intrinsic ferromagnetism, with zero excess energy. A gradient of energy across the N-F_2 interface pushes electrons into F_2, thus lowering the total energy of the electrons in N. In other words, the electrochemical potential of the up-spin subband in N is higher than that of F_2. A low impedance ammeter measures a positive current flow in the collector loop. If the detector is a high impedance voltmeter, the gradient of the electrochemical potential across the N-F_2 interface, equivalent to an electric field, generates a voltage $eV_s = U_M$, positive with respect to ground, at the back of F_2. If M_c is antiparallel to M_e, the energy of N can be lowered by pulling down-spin electrons from F_2 into N, and counterclockwise current is generated in the collector arm of the circuit. An ammeter will register a negative current, and a voltmeter in the collector arm will result in a voltage $-V_s$, negative with respect to ground, observed at the back of F_2.

Thus, positive or negative current (voltage) can be observed by changing the relative orientation of the F_1 and F_2 magnetizations to parallel or antiparallel, which acts as a memory. The V_s is given by

$$\begin{aligned} U_M/e &= \beta M/e\chi \\ &= \beta^2 I_e T_2/e^2\chi\, A\, d \end{aligned} \tag{7.1}$$

This expression can be represented as Eq (7.2) for the free electron model, using $\chi = \beta^2\, N(E_F)$, $\rho = 1/e^2 DN(E_F)$ and $\delta_s = (DT_2)^{1/2}$, where ρ, D, $N(E_F)$ are resistivity, diffusion coefficient and density of states at the Fermi level, respectively [5].

$$V_s = Z_s I_e \tag{7.2}$$

$$Z_s = \rho\, \delta_s{}^2/Ad \tag{7.3}$$

Thus, the voltage V_s is proportional to the emitter current density.

7.1.2.2 Microscopic Transport Model

Simple band models can be used to illustrate device operation [4]. In Fig.7.3 N is depicted with a free electron model and F_1 and F_2 are each depicted with a band model of a transition metal ferromagnet where the two spin subbands of the 3d band are mutually offset by the exchange energy U_x. A fully quantitative microscopic transport model can be used to treat realistic models of a ferromagnet [4]. The transport of current I_c across the F_1-N interface under an applied voltage $-V$ is described by the interfacial conductance,

$$I_c = (g_\uparrow + g_\downarrow)V, \tag{7.4}$$

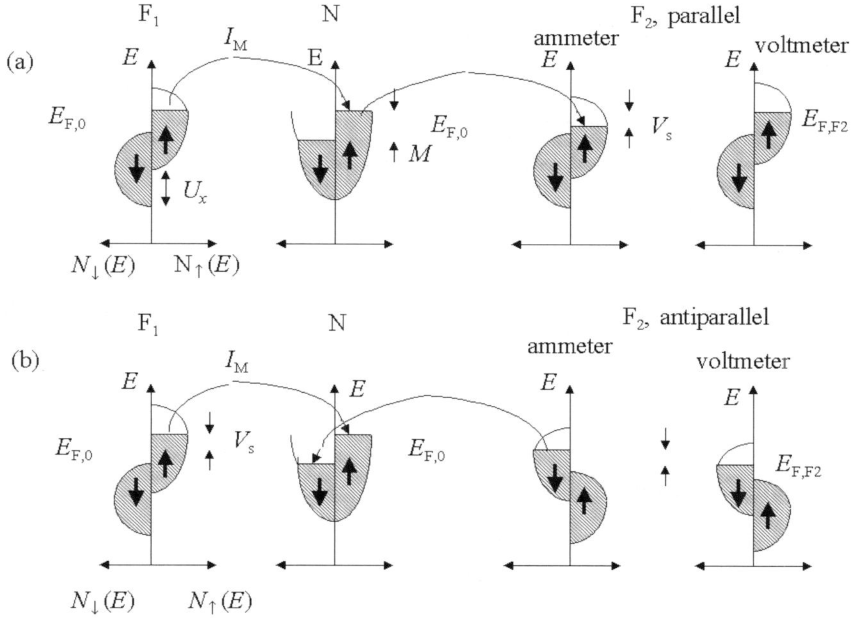

Fig. 7.3 Microscopic models of electron transport, using plots of the density of states, $N(E)$ as functions of energy E for the F_1-N-F_2 system with parallel (**a**) and antiparallel (**b**) configurations of the ferromagnets F_1 and F_2.

where g_s (s =↑or↓) is the conductance per unit area for each spin subband s. The magnetization current of is

$$I_M = \beta\,(g_\uparrow - g_\downarrow)\,V/e. \tag{7.5}$$

The ratio of I_M to I_e

$$I_M / I_c = \beta(g_\uparrow - g_\downarrow)/(g_\uparrow + g_\downarrow)e = \eta\beta/e, \tag{7.6}$$

defines the interfacial spin transport parameter η,

$$I_M = \eta\beta\, I_c/e. \tag{7.7}$$

When M_1 and M_2 are parallel, a positive current is generated across the N-F_2 interface because $g_\uparrow > g_\downarrow$. Equivalently stated, the current from N to F_2 is larger than the back current from F_2 to N. The current across N-F_2 is given by

$$I_c = [\, g_\uparrow(E_{F;N,\uparrow} - E_{F;F2}) + g_\downarrow(E_{F;N,\downarrow} - E_{F;F2})]/e, \tag{7.8}$$

where $E_{F;F2} = E_{F;F2}\uparrow = E_{F;F2}\downarrow$. From $E_{F;N,\uparrow}/E_{F;N}\downarrow = E_{F;N} \pm\beta M/\chi$, (7.8) can be rewritten as

$$I_c = [\,(E_{F;N} - E_{F;F2})\,(g_\uparrow + g_\downarrow) \pm\beta M(g_\uparrow - g_\downarrow)/\chi]\,/\,e. \tag{7.9}$$

Here, $E_{F;N}$ is the average chemical potential of N and the first term is seen to correspond to the nonpolarized current that flows in response to a voltage $(1/e)(\,E_{F;N}$ -

$E_{F;F2}$) across the N-F_2 interface. The second term corresponds to the spin transistor mechanism: a positive current I_c is generated when M_1 and M_2 are parallel and a negative current is generated when M_1 and M_2 are antiparallel. When the device has high impedance there is no current flow, $I_c = 0$, and (7.9) leads to

$$-(E_{F;N} - E_{F;F2})/e = \beta M(g_{\uparrow} - g_{\downarrow})/(g_{\uparrow} + g_{\downarrow})\chi e, \tag{7.10}$$

which is identified as

$$V_c = \eta\beta M/\chi e. \tag{7.11}$$

Comparing (7.11) with (7.1) V_c is reduced by the factor η, and it can be shown that the transimpedance is reduced:

$$Z_s = \eta_e \eta_c \rho_N \delta_s^{\ 2}/Ad. \tag{7.12}$$

7.1.3 Device Properties

From this simple picture we can understand some of the attributes of the spin transistor. (i) Because it is fabricated from metals it is necessarily a low-impedance device. (ii) It is biased by a current, and the bipolar output can be either a current or a voltage. (iii) The magnetization M is linearly proportional to the bias current so that the output characteristics can be described in units of impedance, and the term "spin coupled transimpedance" Z_s is used. (iv) Prototype systems have shown no nonlinearities in their I-V characteristics. This contrasts with the behavior of semiconductor p-n junctions, and implies that the basic, three-terminal embodiment of a spin transistor will not show power gain. (v) The dimensions of magnetization are dipole moments per unit volume. This implies that for a constant bias current the device characteristics (*ie.*, Z_s) can be improved by shrinking the device volume. Note that semiconductor devices can shrink in size until their performance is limited by the dominance of surface states. There is no analog of surface states in metals, so the device dimensions can shrink so long as the current densities do not exceed ~10^8 A/cm^2.

7.1.4 Demonstration of a Prototype Device

Because hysteresis is one of the useful properties of ferromagnetic films, the magnetization M exhibits a memory effect, and we will see below that one of the natural applications of the device is that of a memory element. Fig. 7.4(a) is a hysteresis loop for the device [2]. For fields applied along the z-axis in the film plane the magnetization of the emitter follows the dotted trace and has coercive force $H_{c,e}$, and that of the collector follows the solid trace and has coercive force $H_{c,c}$. Square hysteresis loops similar to those of Fig. 7.4(a) are characteristic of single-domain films that change their orientation over a narrow field range ΔH. Operation of the spin transistor depends on the ability to manipulate the magnetization M individually, and it becomes crucial to use emitter and collector films that have different coercive forces.

A demonstration of device operation, in voltage mode, is presented in Fig. 7.4(b). A constant bias current is supplied and the voltage from the collector to ground is recorded as the magnetization and direction of external field is changed. The detected voltage was linear with bias current and is displayed in units of impedance. The solid line is a trace for which the field was swept from an initial negative value to a final positive value. The recorded voltage is positive for fields $H < H_c$ because M_c and M_e are aligned parallel along $-z$. When H is slightly larger than $H_{c,e}$ but smaller than $H_{c,c}$, the voltage becomes negative because M_c and M_e are aligned antiparallel. When H is increased above $H_{c,c}$ the original positive voltage is recovered because M_c and M_e are aligned parallel again, now along $+z$. The dotted line is for a reverse trace. The data in Fig. 7.4(b) were taken at cryogenic temperatures (4 K < T < 90 K). There is no physical reason to prevent room temperature operation of the device. Because of hysteresis these devices exhibit a memory effect, an example of which is shown in Fig. 7.4(c). Thus, the spin transistor is expected as a nonvolatile RAM with high density. However, the output voltage is small of order μV at most.

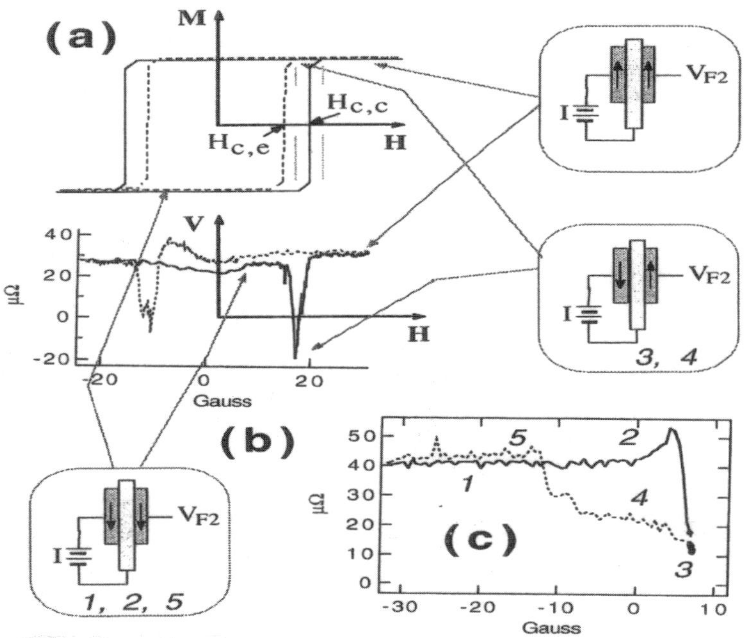

Fig. 7.4 (**a**) Hysteresis loops for ferromagnetic emitter and collector films. (**b**) Examples of data from a prototype. (**c**) Example of the memory effect.

7.2 Spin FET

A device that has been proposed but not yet demonstrated applies the spin injection concept to a semiconductor structure, yielding a spin-polarized field effect transistor (spin FET [6]). Figure 7.5 is a schematic diagram of a spin FET. The current-carrying medium would be an inversion layer formed at the heterojunction between InAlAs and InGaAs. The two-dimensional electron gas (2DEG) in that layer would provide a very high-mobility channel, free of spin-flip scattering events. The Fermi wave vector $k_F = (2\pi n_s)^{1/2}$ of 2DEG carriers is of order 10^6 cm^{-1} for typical carrier densities of order $n_s \sim 10^{12}$ cm^{-2}, and the Fermi velocity $v_F \sim 10^7$ cm/s is weakly relativistic. The spin-polarized carriers are injected and collected by ferromagnetic metal pads. However, one can expect a strong internal electric field E_z to be present in the heterostructure interface region in the inversion layer, oriented perpendicular to the layer. This field induces an interface spin-orbit effect on the carriers in the channel moving parallel to the interface. When the source of E_z is a gradient of the confining potential, the appropriate spin-orbit Hamiltonian is

$$H_{SO} = \alpha(\sigma \text{ x } \boldsymbol{k}) \bullet z, \tag{7.13}$$

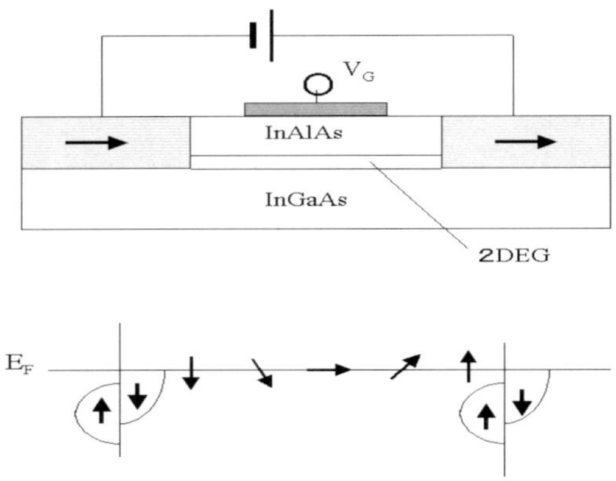

Fig. 7.5 Spin-polarized field-effect transistor scheme. V_G is the gate voltage

where σ are the Pauli matrices and α describes the strength of the spin-orbit coupling. The spin splitting energy $\Delta_{SO} \sim 2\alpha k_F$ = 2 to 5 meV have been measured by analyzing the beat pattern of Shubnikof–De Haas oscillations [7]. The field will cause the spins of the carriers to precess. This precession will rotate them out of alignment with the magnetization of the second ferromagnetic pad, decreasing the transmitted current of the device. Finally, if a gate electrode is deposited on top of the device, one can apply a gate voltage V_G to increase or decrease the effective electric field, altering the spin precession. This will control the alignment of the carriers' spin with respect to the magnetization vector in the second pad, thus permitting modulation of the current passing through the device. The carriers' spin precession angle is given by [6,8]

$$\Delta\theta = 8\pi^2 m^* \alpha L/h^2, \tag{7.14}$$

$$\alpha = \langle E\rangle (h^2/8\,\pi^2 m^*)\Delta_{SO}\,(2E_g+\Delta_{SO})/E_g(E_g+\Delta_{SO})(3E_g+2\Delta_{SO}), \tag{7.15}$$

where $\langle E\rangle$ is the expectation value of the electric field, m^* the effective mass of the electrons, L the distance between the ferromagnetic pads, E_g the band gap of the semiconductor and Δ_{SO} the spin splitting energy in the valence band due to spin-orbit interaction.

The main research subjects for realizing a spin FET are as follows;

1. modifying α by the gate voltage
2. spin injection from the ferromagnetic pad to the semiconductor
3. transport behavior of spin polarized electrons in the 2DEG
4. operation at RT.

The first has recently been verified by Nitta *et al* [7]. The second remains unclear, although there are some papers describing the success of spin injection to semiconductors. With respect to the third problem it has been observed that the spin diffusion length in the 2DEG is of order several μm, and an electric field could push the spin polarized carriers over distances of order 100 μm [9,10]. The operation of a spin FET at RT needs small band gap semiconductors as suggested by (7.14) and (7.15), because δ_{SO} is as small as several meV.

7.3 Spin-Valve Transistor

7.3.1 Principle of the Spin-Valve Transistor

The perpendicular electron transport and exponential mean free path dependence in a metal base transistor allows for fundamental detection of the perpendicular spin valve effect by incorporating a spin valve into the base. A spin-valve transistor (SVT) consists of spin-valve multilayers sandwiched between two potential barriers made of two semiconductors (SMS structure) or an insulator and a semiconductor (MOMS structure) [11]. The emitter barrier injects hot electrons into the spin valve base and the collector barrier accepts only ballistic electrons, making the collector current very sensitive to magnetic fields. In the case of the SMS structure the Schottky injector allows emitter current densities sufficiently high for room temperature operation of the spin-valve transistor, compared with the MOSM structure, and is therefore preferred. The energy band diagram of the Co/Cu spin valve transistor is shown in Fig. 7.6 [11]. The base is formed by a spin-valve. A Co/Cu/Co/Pt sandwich base is sputtered onto a Si(100) collector substrate. The collector (emitter) barrier height is about 0.6 eV (0.7 eV). The emitter Schottky barrier is slightly higher than that of the collector, because of quantum mechanical reflections. The emitter and collector Schottky barriers are in forward and reverse bias, respectively. The emitter bias accelerates the electrons over the emitter barrier, after which they constitute the hot, ballistic electrons in the base. The probability of passing the collector barrier is limited by collisions in the base, which depend on the magnetic state of the spin-valve layer, by scattering in semiconductors and quantum mechanical reflections at the base-collector interface. A Pt capping layer on top of the spin valve is used to make the emitter Schottky barrier larger than the collector barrier, in order to decrease quantum mechanical reflections at the collector barrier. For a metal base transistor with a single metal base film this can be expressed by the common base current transfer ratio or current gain $\alpha_0 = (J_c - J_{leak})/J_e$, where J_c is the total collector current, J_{leak} is the collector leakage current, determined by the reverse-biased collector Schottky barrier and J_e is the injected emitter current. The spin valve transistors sensitivity to the electron mean free path instead of the macroscopic resistance allows more direct access to the physical parameters (scattering length) in the spin-valve effect. Most interestingly, it provides a way to probe the spin dependent scattering mechanism as a function of electron energy.

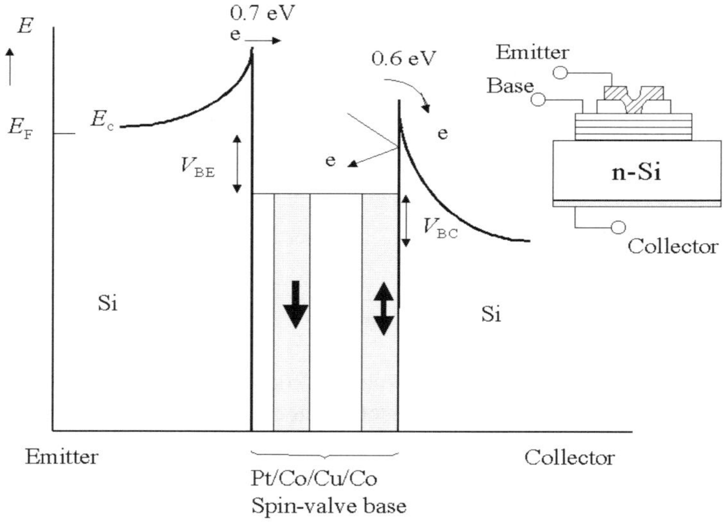

Fig. 7.6 Energy band diagram of a Co/Cu spin-valve transistor

The magnetocurrent, namely the relative change in collector current $MC = (I_C^P - I_C^{AP})/I_C^{AP}$, is given by (7.16), where I_C^P and I_C^{AP} are the collector current in the parallel and antiparallel alignment of magnetizations of the base, respectively:

$$MC = \cosh (W(1-\lambda_\uparrow/ \lambda_\downarrow)/2\, \lambda_\uparrow) - 1, \qquad (7.16)$$

where W is the base thickness and $\lambda_\uparrow$ and $\lambda_\downarrow$ are the mean of the injected hot electrons with $\uparrow$and $\downarrow$spins in the base, respectively.

7.3.2 Device Properties

Excellent quality semiconductor materials are important for good Schottky barriers and proper ballistic transmission of hot electrons. In general, deposition of semiconductors on metal leads to severe interdiffusion of species even at RT and show bad semiconducting characteristics, such as mobility and number of defects, resulting in poor electrical characteristics. Hence, deposition of semiconductors on the base metal is not appropriate to acquire reliable devices. The dilemma of growing a device-quality semiconductor on top of the metallic layer was tackled by sandwiching the base between two crystalline substrates [12]. The so-called direct bonding technology provided a spontaneous adhesion based on Van der Waals forces and hydrogen bonds. Using this structure a large perpendicular hot-electron spin valve effect (400% at 77 K) has been observed for electrons of about 0.6 eV above the Fermi level in Co-Cu [12]. Experiments at 77 K were conducted to decrease the collector leak current with respect to the hot electron current.

Room temperature operation requires reduction of the device area, increase of the collector barrier height and enlargement of the base transport factor. A further direction is a sufficiently large injection current density, which could only be obtained using a Schottky barrier injector. Recently, a SVT structure was developed with very large magnetotransport effect at RT [13]. The output current of this device changes by more than a factor of three in magnetic fields of only a few Oe, corresponding to a magnetocurrent above 200% and field sensitivity of 130% per Oe. This substantial improvement in sensitivity and low driving field was achieved by (1) selection of suitable metal layers for the emitter and collector Schottky diodes, (2) optimizing the processing to realize well-defined Schottky diodes with extremely low leakage current and (3) the introduction of a soft spin valve of Pt/NiFe/Au/Co/Au as the base layer. On one side of the spin valve an additional Pt layer is incorporated to have a Si-Pt interface which has a high Schottky barrier (0.86 ±0.01 eV). On the other side of the spin valve an additional Au layer is incorporated to form a lower Si-Au Schottky barrier (0.75 ± 0.01 eV).

At the emitter and collector side of the SVT, n-type Si(100) and Si(111) are used, respectively. The bonding was done between 2 nm Au on the spin-valve side and 2 nm of Au at the collector side, to obtain the structure Si(100)/Pt(2 nm)/NiFe(3 nm)/Au(3.5 nm)/Co(3 nm)/Au(2 nm;2 nm)/Si(111). This bonded structure was processed into transistors with a 350 x 350 μm^2 emitter area and a 700 x 350 •m^2 collector area. In Fig. 7.7 the magnetic field dependence of the collector at RT (295 K) as well as at 83 K is given for an emitter current $I_E = 2$ mA [13]. The collector diode is zero biased ($V_{BC} = 0$). It is clearly seen that a large change in the collector current is obtained by a small change in the magnetic field, not only at low temperature but also at RT. At RT a huge MC of 208% is realized and at 83 K, MC equals 402 %. The onset of the switching field of NiFe is at ~4 Oe and the switching is complete at ~ 7 Oe. The inset in the figure shows the current-in-plane magnetoresistance of the spin valve incorporated in the SVT. This shows a CIP magnetoresistance of only 0.5% at RT.

Obviously, a very large relative effect is not the only factor that determines the application potential. At present the collector current in this SVT is very low, which is caused by the low transfer ratio of the device (10^{-6}). This may lead to a

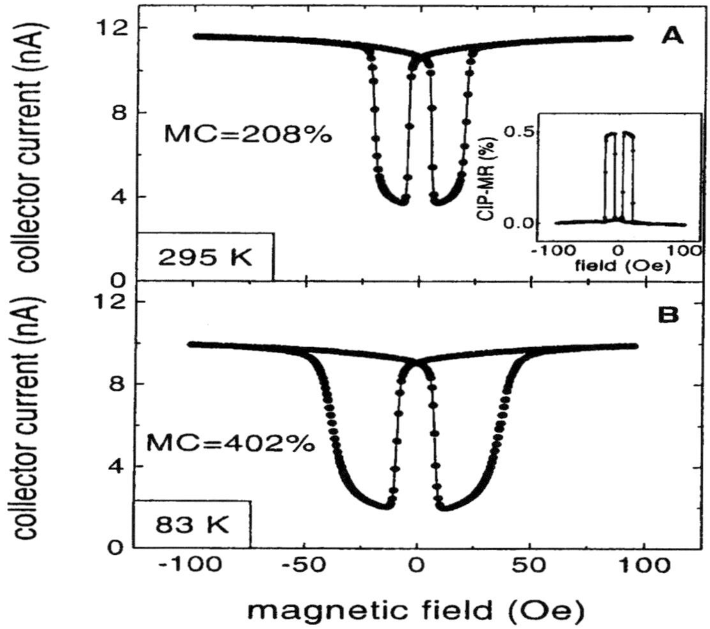

Fig. 7.7 Magnetic field dependence of collector current of the spin-valve transistor (**A**) at RT showing a spin current of 208% and (**B**) at 83 K showing a 402% spin current. The inset in A shows the RT current in plane MR of the spin valve.

poor signal to noise ratio for high-speed operation. Effort must be made to improve the output current and characterize the noise behavior.

7.4 Spin-Dependent Single-Electron Tunneling Devices

7.4.1 Structures and Properties

Electron tunneling into a small particle (dot or cluster) gives rise to an interesting effect, namely the Coulomb blockade. Tunneling of an electron into a small cluster increases the energy of the system by a charging (Coulomb) term, $E_C = e^2/2C$, where C is the capacitance of the cluster. This charging energy introduces a gap (Coulomb gap) for electron tunneling and gives rise to the Coulomb blockade effect. If the applied voltage V and the thermal energy $k_B T$ are much smaller than E_C, only electrons at E_C above the Fermi energy can tunnel, thus increasing the tunnel resistivity by a factor of the order of $\exp(E_C/2k_BT)$. When eV or $k_B T$ exceeds E_C, the base tunnel resistance is approximately restored. Thus, the I-V curve for electron tunneling accompanied by the Coulomb blockade effect has current steps known as the Coulomb staircase, which is shown in Fig. 7.8.

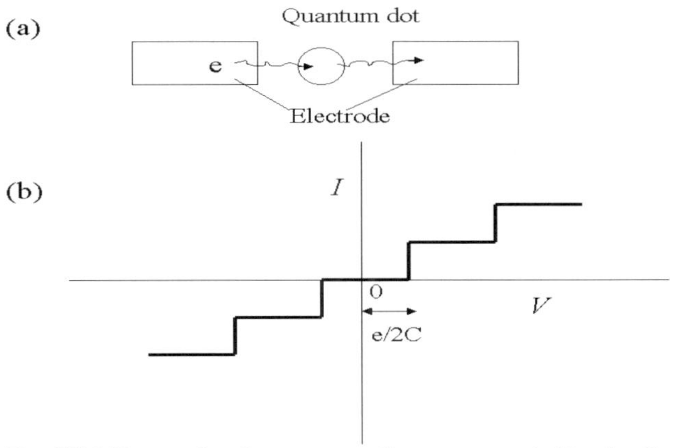

Fig. 7.8 I-V curve for electron tunneling accompanied by the Coulomb blockade effect

The coexistence of spin-dependent tunneling and Coulomb blockade was first seen for Ni clusters in SiO_2 [14] and more recently investigated by Fujimori *et al.* [15] for several types of granular systems, of which the model structure is shown in Fig. 7.9. When the cluster concentration is just below the percolation threshold, the conduction is dominated by tunneling between metallic clusters, with a tunnel resistance enhanced by the Coulomb blockade at low temperatures and dependent on the relative orientation of the magnetization in the two clusters (it decreases when an applied field aligns the magnetic moments of the clusters). However, in granular materials the conduction involves a large number of clusters, with generally a broad distribution of size and intercluster distance. This leads to relatively complex Coulomb blockade effects that cannot be analyzed as quantitatively as in the simpler structures usually fabricated for Coulomb blockade studies [14,15,16].

The TMR in granular system is given by the formula $P^2/(1+P^2)$, which is half of that for magnetic tunneling junctions because of the difference between random and antiparallel alignment of magnetic moments. Typical electrical resistivity and magnetization as a function of magnetic field are shown at RT for a $Co_{52}Al_{20}O_{28}$ film in Fig. 7.10 [17]. The magnetization versus magnetic field curve can be fitted by the Langevin function, implying that these films are superparamagnetic. Comparing the ρ versus H curve and the M versus H curve the relation $\Delta\rho/\rho_0 \propto -(M/M_s)^2$ can be obtained as shown in Fig. 7.10, where M_s is the value of M at H = 50 kOe, $\Delta\rho = \rho(H) - \rho_0$ and ρ_0 is ρ at $H = 0$. This means that the change in ρ by H takes place through the alignment of the magnetizations of granules, which is proportional to the square of the averaged magnetization.

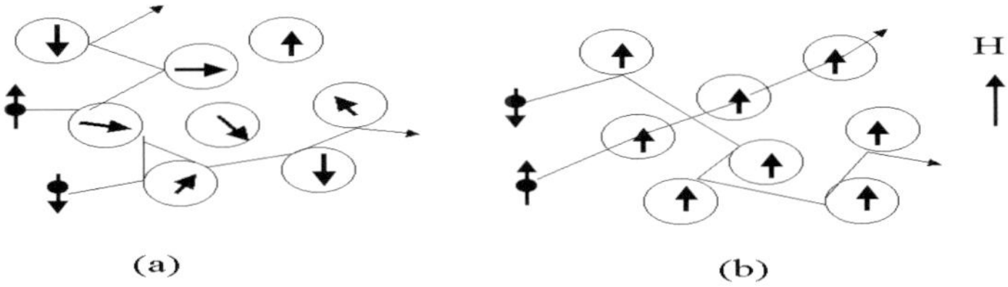

Fig. 7.9 Schematic diagram for spin-dependent tunneling in the granular system.

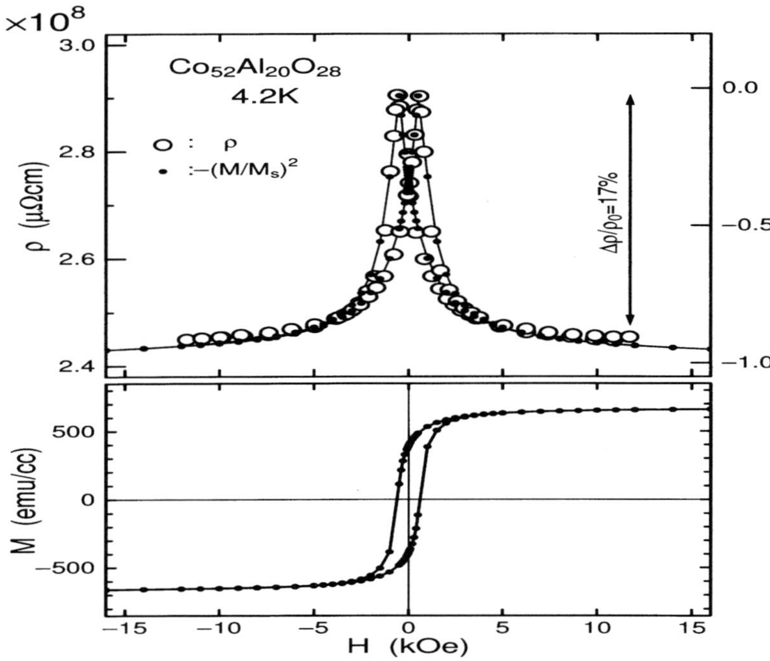

Fig. 7.10 TMR curve for the Co-Al_2O_3 granular system.

The relatively small field sensitivity of MR in granular systems was improved by a hybrid structure, called nanogranular in gap (GIG) films [18]. In the structure granular film is placed in the lateral gap between a couple of soft magnetic films such as Permaloy, which is shown in Fig.7.11 schematically. The GIG was fabri-

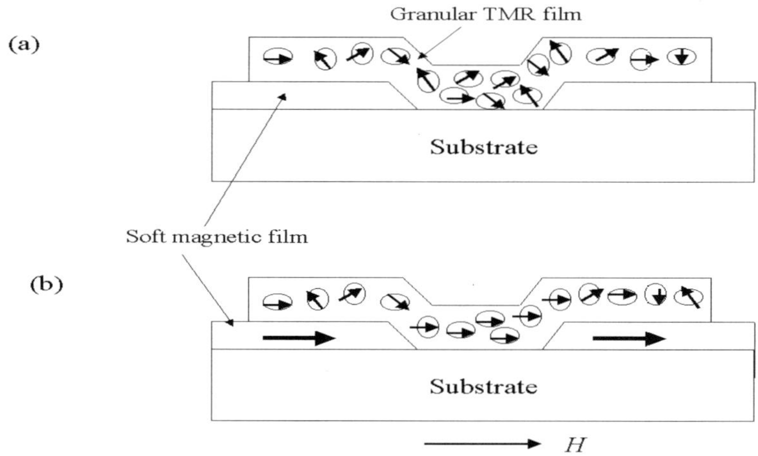

Fig. 7.11 Schematic illustration of the GIG film in (**a**) zero field and (**b**) an applied field

cated by a microprocessing technique using ion milling and rf sputtering methods. First, a ~2μm thick $Fe_{66}Ni_{34}$ film was deposited onto a glass substrate by rf sputtering using a $Fe_{66}Ni_{34}$ alloy target, and then annealed at 300° C for 2 hours in vacuum. Second, a gap of about 5 μm width was made in the film by photolithography and ion milling etching. Third, a ~1 μm thick film was deposited onto the $Fe_{66}Ni_{34}$ film, in the gap region by rf sputtering using a target composed of MgF_2 chips placed on a CoFe disk. In this structure the large field from the soft magnet magnetized by a small external field can be provide to the granular film, thus the field sensitivity of the MR is expected to be improved. Figure 7.12 shows the MR curve of the GIG film in a low magnetic field. The low field sensitivity of the MR is remarkably improved in the GIG film.

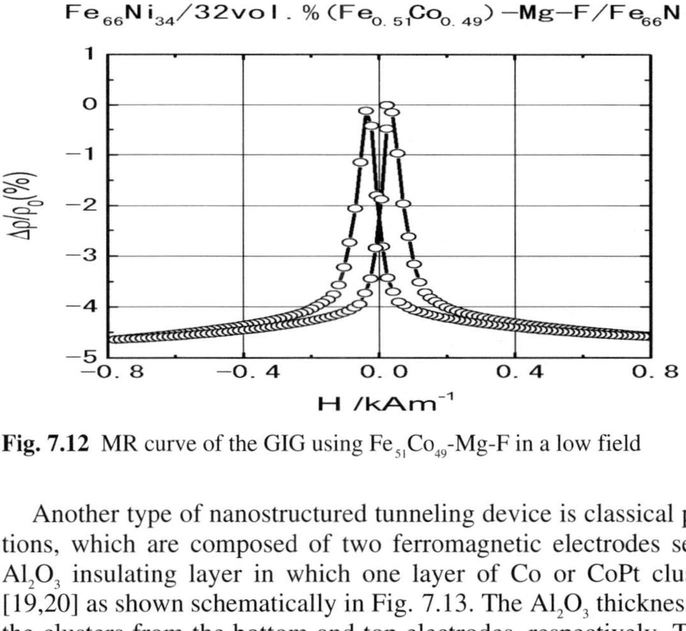

Fig. 7.12 MR curve of the GIG using $Fe_{51}Co_{49}$-Mg-F in a low field

Another type of nanostructured tunneling device is classical planar tunnel junctions, which are composed of two ferromagnetic electrodes separated by a thin Al_2O_3 insulating layer in which one layer of Co or CoPt clusters is embedded [19,20] as shown schematically in Fig. 7.13. The Al_2O_3 thickness d and d' separate the clusters from the bottom and top electrodes, respectively. The clusters are superparamagnetic in the range 1– 4 nm at higher temperatures, while they become

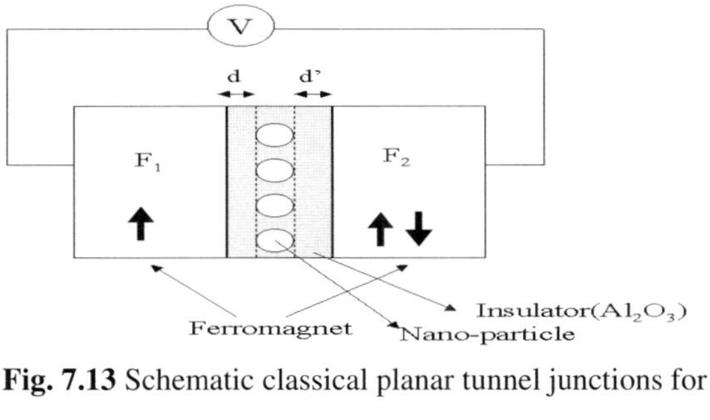

Fig. 7.13 Schematic classical planar tunnel junctions for a nano-structured tunneling device

ferromagnetic for the larger cluster size, in particular, for CoPt clusters due to the large magnetocrystalline anisotropy of the CoPt alloy [20]. TMR can be expected by changing the relative orientations of magnetizations between the magnetic electrodes and the clusters. This type of planar junctions is suitable to investigate spin-dependent tunneling accompanied with Coulomb blockade effects.

A cross sectional transmission electron microscopy image for a classical planar tunnel junction is shown in Fig. 7.14, which was prepared by sputtering and has layered CoPt nanoparticles with average diameter of 10 nm separated by a thin Al_2O_3 insulating layer between two ferromagnetic electrodes. This double junction with a discontinuous intermediate layer exhibits large TMR at RT due to the large coercive force of the CoPt nanoparticle due to the large magnetocrystalline anisotropy as shown in Fig. 7.15 [21]. Fe/CoFe double layers were used for both the

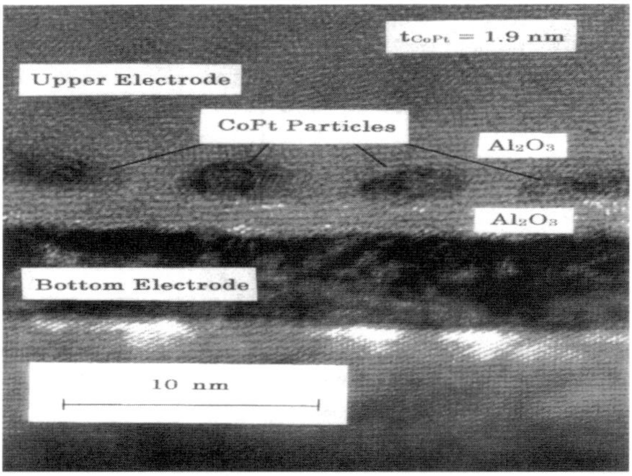

Fig. 7.14 Cross-sectional transmission electron microscopy image for a classical planar tunnel junction

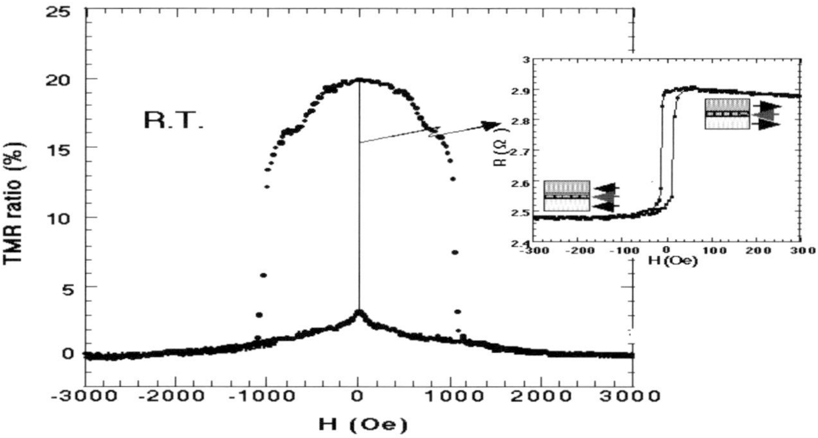

Fig. 7.15 TMR curve at RT for nanostructured spin tunneling junctions with large coercive force of the CoPt nanoparticle

electrodes in the double junction, which have lower coercive force than that of the CoPt nanoparticles. Thus, the MR attributes to the coercive force differences between the electrodes and the nanoparticles.

Nanostructured spin tunneling double junctions having a discontinuous intermediate layer can be fabricated with a low barrier height without geometrical enhancement of TMR, which occurs in the cross-shaped mm^2-size tunneling junctions with low barrier height [22]. Figure 7.16 exhibits the temperature dependence of the TMR for the junctions with a barrier height of 0.4 eV [20], which demonstrates that the TMR with a low barrier height is almost constant in a wide *T* range between 5 K and 300 K. This temperature-independent TMR is attributed to the broadening of the Fermi distributions. When the charging effect does not exist, the total conductance $G(\theta, T)$ and conductance change $\Delta G(T)$ are expressed as [23]

$$G(\theta, T) = G_T(T)\,[1 + P_1(T)P_2(T)\cos(\theta)] + G_i(T), \tag{7.17}$$

$$\Delta G(T) = 2G_T(T)\,P_1(T)P_2(T), \tag{7.18}$$

where θ is the angle between the magnetization directions, and $G_T(T)$ and $G_i(T)$ are the elastic tunneling due to the broadening of the Fermi distributions and the conductance which is independent of the relative orientation of the electrode magnetization, respectively. The $\Delta G(T)$ was observed to increase in nanostructured double junctions, which cannot be attributed to the *T* dependence of spin polarization, since it has the wrong sign. Thus, the temperature-independent TMR is ascribed to the broadening of the Fermi distributions, which becomes remarkable for junctions with low barrier height.

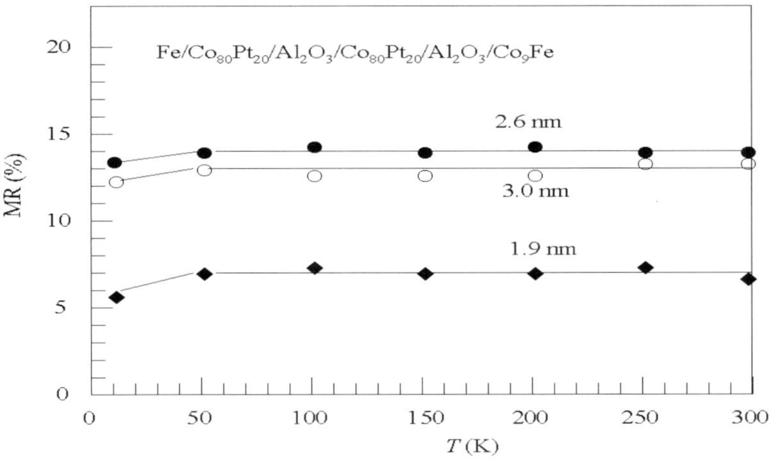

Figure 7.16 Temperature dependence of the TMR in $(Fe/Co_{80}Pt_{20})/\ Al_2O_3/\ Co_{80}Pt_{20}$ (t nm)/$Al_2O_3/Co_{90}Fe_{10}$ nanostructured spin tunneling junctions with a low barrier height of 0.4 eV. Solid lines are to guide the eye

7.4.2 TMR Enhancement by Coulomb Blockade Effect

The anomalous increase of TMR at low temperatures has been observed in the Co-Al-O granular film system [24] as shown in Fig. 7.17, in which the MR is remarkably enhanced at low temperatures while it is nearly independent of temperature above ~ 100 K. For $Co_{30}Al_{22}O_{42}$, the MR below 3 K is anomalously large and reaches more than twice the value given by $P^2_{Co}/(1+P^2_{Co})$, where the formula is half that for magnetic tunneling junctions because of the difference between random and antiparallel alignment of magnetic moments. The enhancement of the TMR in granular systems has been ascribed to cotunneling in the Coulomb blockade regime, which is of a higher order in the tunneling resistance and so the difference in the spin-dependent tunneling between the antiparallel and the parallel configuration is larger [25, 26]. The enhancement of TMR due to cotunneling was first observed in microfabricated ferromagnetic single-electron transistors [26].

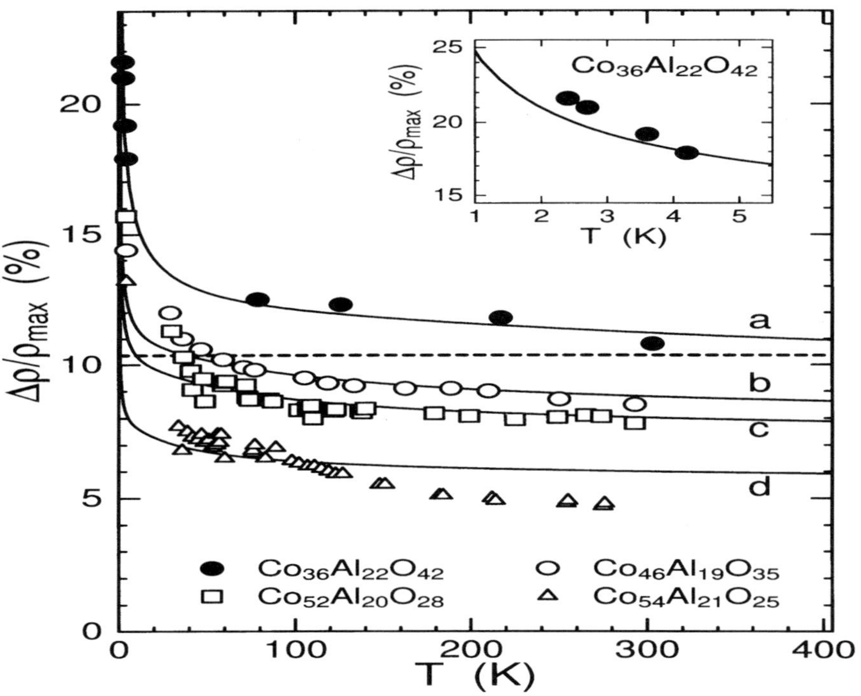

Fig. 7.17 Temperature dependence of MR for Co-Al-O granular films

Theoretically charging effects have been investigated for the simplest structure of single-electron tunneling consisting of two electrodes and an island, which is separated from the electrodes by tunnel barriers (double junction). When the two electrodes and the island are ferromagnetic, then charging effects lead to oscillatory-like behavior of TMR versus applied voltage [27, 28]. In the theories the limit of sequential tunneling and the limit of fast intrinsic spin relaxation time on the island (the relaxation time is shorter than or of the order of the time between successive tunneling events) were considered. In this limit there is no TMR when the island is nonmagnetic. To obtain finite TMR it is necessary to admit spin accumulation on the island by assuming a long spin relaxation time (much longer than the time between two successive tunneling events).

The clear TMR oscillation as a function of the applied voltage has been observed at low temperatures, accompanied with the conductance oscillation in the nanostructured double tunneling junctions as shown in Fig. 7.14 with smaller CoPt particles being superparamagnetic at RT [29]. This observation is shown in Fig. 7.18. This exhibits an oscillation with a period of 1.6 mV, which was attributed to the single-electron charging effect [29]. The small oscillation period, however, is rather curious considering the charging effect. Recently clearer TMR oscillation due to the charging effect was observed for microfabricated nanostructured tunneling junctions [30]. This tunneling junction exhibits the Coulomb staircase for the I-V curve. The oscillation period coincides with the Coulomb gap. Namely, TMR enhancement appears through nonlinear tunneling due to the charging effect.

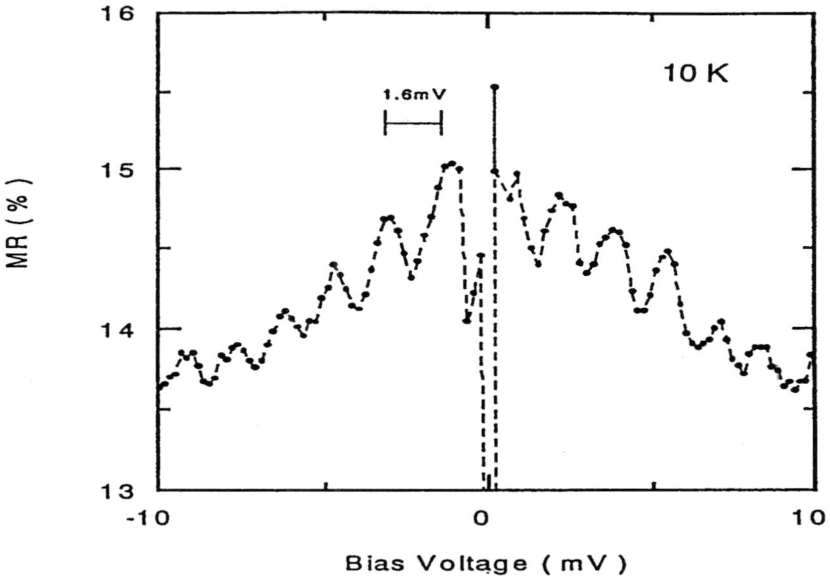

Fig. 7.18 Bias voltage dependence of the tunneling magnetoresistance and the tunneling conductance measured at 10 K

7.4.3 Spin Accumulation

In a ferromagnetic single-electron transistor and nanostructured double tunneling junctions, the transition rates for tunneling into or out of the island depend on the electron spin. An electric current through the system therefore implies a spin current out of or into the island. This creates a nonequilibrium excess spin on the island. The excess spin decays by spin flip relaxation. The nonequilibrium spin accumulation on the island is equivalent to a nonequilibrium chemical potential difference between the spin-up and the spin-down states. The TMR effect is then related to the resulting spin split of the chemical potential and this split is governed by the balance between the spin-flip rate on the island and the spin-flip rate due to electron tunneling into and out the island. The nonequilibrium spin is important for the transport properties of the ferromagnetic single-electron transistor and nanostructured double tunneling junctions when the nonequilibrium chemical potential difference $\Delta\mu$ is of the same order as the Coulomb charging energy E_c. Theoretical studies predict that first, spin accumulation in a nonmagnetic island can generate TMR, which disappears in the limit of fast intrinsic spin relaxation on the island and saturates for spin relation time [27, 28]. Second, the spin polarization of the island and TMR have oscillatory and nonoscillatory components. The oscillations as a function of the voltage result from the interplay of the Coulomb energy and spin-dependent shifts of the chemical level, which are induced by spin accumulation. The oscillation due to spin accumulation in a nonmagnetic island, however, has not yet been observed experimentally.

References

[1] G. A. Prinz: *Science*, **250**, 1092 (1990).
[2] M. Johnson: *J. Magn. Magn. Mater.* **156**, 321 (1996)
[3] R. B. Lewis and T. R. Carrer: *Phys. Rev.* **155**, 309 (1967)
[4] M. Johnson: *J. Mag. Soc. Jpn.* **19**, 684 (1995).
[5] M. Johnson: *J. Magn. Magn. Mater.* **140-144**, 21 (1995)
[6] S. Datta and B. Das: *Appl. Phys. Lett.*, **50**, 665 (1990).
[7] J. Nitta, T. Akazaki and H. Takayanagi; Phys. Rev. Lett. **78**, 1335 (1997).
[8] E. A. de Andrada E Silava, G. C. La. Rocca and F. Bassari: *Phys. Rev.* **B50**, 8523 (1994).
[9] J. M. Kikkawa and D. D. Awschalom: *Phys. Rev. Lett.*, **80**, 4313 (1998).
[10] D. Hagelle, M. Ostreich, W. Ruhle, N. Nestle and K. Ebels: *Appl. Phys. Lett.* **73**, 1580 (1998).
[11] D. J. Monsma, J.C. Lodder, Th. J. A. Pompa, B. Dieny: *Phys. Rev. Lett.* **74**, 5260 (1995).
[12] D. J. Monsma, R. Vlutters, J.C. Lodder: *Science* **281**, 407 (1998).
[13] P. S. A. Kumar, R. Jansen, D. M. J. van't Erve, R. Vlutteres, P. de haan, J. C. Lodder: *J. Magn. Magn. Mater.* **214**, L1 (2000).
[14] J. S. Helman and B. Abels: *Phys. Rev. Lett.* 37, 1429 (1976).
[15] H. Fujimori, S. Mitani and S. Ohnuma: *Mater. Sci. Engn.* **B31**, 219 (1995).
[16] A. Milner, A. Gerber, B. Groisman, M. Karpovsky and A. Gladkikh: *Phys. Rev. Lett.* **76**, 475 (1996).
[17] S. Mitani, H. Fujimori and S. Ohnuma: *J. Magn. Magn. Mater.* **165**, 141 (1997).
[18] N. Kobayashi, S. Ohnuma, S. Murakami, T. Masumoto, S. Mitani and H. Fujimori: *J. Magn. Magn. Magn. Mater.* **188**, 30 (1998).

[19] L. F. Shep, A. Fert, F. Fettar, P. Holody, S. F. Lee, J. L. Maurice, P. Petroff and A. Vaures: *Phys. Rev.* **B56**, R5747 (1997).

[20] K. Inomata, H. Ogiwara, Y. Saito, K. Yusu and K. Ichihara: *Jpn. J. Appl Phys.* **36**, L1380 (1997). , K. Inomata and Y. Saito: *Appl. Phys. Lett.* **73**, 1143 (1998).

[21] Y. Saito, K. Nakajima, K. Tanaka and K. Inomata: *IEEE Trans. Magn.* **35**, 2904 (1999).

[22] J. S. Moodera, L. R. Kinder, J. Nowark, P. Leclair and R. Meservey: *Appl. Phys. Lett.* **69**, 708 (1996).

[23] C.H.Cheng, J. Nowark, R. Jansen and J.S. Moodera: *Phys. Rev.* **B58**, R2917 (1998).

[24] S. Mitani, T. Takahashi, K. Takanashi, K. Yakushiji, S. Maekawa and H Fujimori, *Phys. Rev. Lett.* **81**, 2799 (1998).

[25] S. Takahashi and S. Maekawa: *Phys. Rev. Lett.* **80**, 1758 (1998).

[26] K. Ono, H. Shimada and Y. Ootsuka: *J. Phys. Soc. Jpn.* **66**, 1261 (1997).

[27] J. Barnas and A. Fert: *Phys. Rev. Lett.* **80**, 1058 (1998).

[28] A. Brataas, Y. V. Nazarov, J. Inoue and G. E. W. Baner: *Phys. Rev.* **59,** 93 (1999).

[29] K. Nakajima, Y. Saito, S. Nakamura and K. Inomata, *IEEE Trans. Magn.* **36**, 2806 (2000) .

[30] K. Takanashi, private communication

SPRINGER SERIES IN SURFACE SCIENCES

Editors: G. Ertl, R. Gomer and D.L. Mills

Founding Editor: H.K.V. Lotsch

1 **Physisorption Kinetics**
By H. J. Kreuzer, Z. W. Gortel

2 **The Structure of Surfaces**
Editors: M. A. Van Hove, S. Y. Tong

3 **Dynamical Phenomena at Surfaces, Interfaces and Superlattices**
Editors: F. Nizzoli, K.-H. Rieder, R. F. Willis

4 **Desorption Induced by Electronic Transitions, DIET II**
Editors: W. Brenig, D. Menzel

5 **Chemistry and Physics of Solid Surfaces VI**
Editors: R. Vanselow, R. Howe

6 **Low-Energy Electron Diffraction**
Experiment, Theory
and Surface Structure Determination
By M. A. Van Hove, W. H. Weinberg, C.-M. Chan

7 **Electronic Phenomena in Adsorption and Catalysis**
By V. F. Kiselev, O. V. Krylov

8 **Kinetics of Interface Reactions**
Editors: M. Grunze, H. J. Kreuzer

9 **Adsorption and Catalysis on Transition Metals and Their Oxides**
By V. F. Kiselev, O. V. Krylov

10 **Chemistry and Physics of Solid Surfaces VII**
Editors: R. Vanselow, R. Howe

11 **The Structure of Surfaces II**
Editors: J. F. van der Veen, M. A. Van Hove

12 **Diffusion at Interfaces: Microscopic Concepts**
Editors: M. Grunze, H. J. Kreuzer, J. J. Weimer

13 **Desorption Induced by Electronic Transitions, DIET III**
Editors: R. H. Stulen, M. L. Knotek

14 **Solvay Conference on Surface Science**
Editor: F. W. de Wette

15 **Surfaces and Interfaces of Solids**
By H. Lüth*)

16 **Atomic and Electronic Structure of Surfaces**
Theoretical Foundations
By M. Lannoo, P. Friedel

17 **Adhesion and Friction**
Editors: M. Grunze, H. J. Kreuzer

18 **Auger Spectroscopy and Electronic Structure**
Editors: G. Cubiotti, G. Mondio, K. Wandelt

19 **Desorption Induced by Electronic Transitions, DIET IV**
Editors: G. Betz, P. Varga

20 **Scanning Tunneling Microscopy I**
General Principles and Applications to Clean and Adsorbate-Covered Surfaces
Editors: H.-J. Güntherodt, R. Wiesendanger
2nd Edition

21 **Surface Phonons**
Editors: W. Kress, F. W. de Wette

22 **Chemistry and Physics of Solid Surfaces VIII**
Editors: R. Vanselow, R. Howe

23 **Surface Analysis Methods in Materials Science**
Editors: D. J. O'Connor, B. A. Sexton, R. St. C. Smart

24 **The Structure of Surfaces III**
Editors: S. Y. Tong, M. A. Van Hove,
K. Takayanagi, X. D. Xie

25 **NEXAFS Spectroscopy**
By J. Stöhr

26 **Semiconductor Surfaces and Interfaces**
By W. Mönch
3rd Edition

27 **Helium Atom Scattering from Surfaces**
Editor: E. Hulpke

28 **Scanning Tunneling Microscopy II**
Further Applications
and Related Scanning Techniques
Editors: R. Wiesendanger, H.-J. Güntherodt
2nd Edition

29 **Scanning Tunneling Microscopy III**
Theory of STM
and Related Scanning Probe Methods
Editors: R. Wiesendanger, H.-J. Güntherodt
2nd Edition

30 **Concepts in Surface Physics**
By M. C. Desjonquères, D. Spanjaard*)

31 **Desorption Induced by Electronic Transitions, DIET V**
Editors: A. R. Burns, E. B. Stechel, D. R. Jennison

32 **Scanning Tunneling Microscopy and Its Applications**
By C. Bai
2nd Edition

33 **Adsorption on Ordered Surfaces of Ionic Solids and Thin Films**
Editors: H.-J. Freund, E. Umbach

34 **Surface Reactions**
Editor: R. J. Madix

35 **Applications of Synchrotron Radiation**
High-Resolution Studies of Molecules
and Molecular Adsorbates on Surfaces
Editor: W. Eberhardt

36 **Kinetics of Metal-Gas Interactions at Low Temperatures: Hydriding, Oxidation, Poisoning**
By E. Fromm

37 **Magnetic Multilayers and Giant Magnetoresistance**
Fundamentals and Industrial Applications
Editor: U. Hartmann*)

*) Available as a textbook

Printing (Computer to Film): Saladruck Berlin
Binding: Stürtz AG, Würzburg